CHANCE, DEVELOPMENT, AND AGING

CHANCE,

DEVELOPMENT,

AND AGING

Caleb E. Finch & Thomas B. L. Kirkwood

New York   Oxford
Oxford University Press
2000

Oxford University Press

Oxford   New York
Athens   Auckland   Bangkok   Bogotá   Buenos Aires   Calcutta
Cape Town   Chennai   Dar es Salaam   Delhi   Florence   Hong Kong   Istanbul
Karachi   Kuala Lumpur   Madrid   Melbourne   Mexico City   Mumbai
Nairobi   Paris   São Paulo   Singapore   Taipei   Tokyo   Toronto   Warsaw

and associated companies in
Berlin   Ibadan

Published by Oxford University Press, Inc.
198 Madison Avenue, New York, New York 10016

Oxford is a registered trademark of Oxford University Press

Library of Congress Cataloging-in-Publication Data
Finch, Caleb Ellicott.
Chance, development, and aging / by Caleb E. Finch, Thomas B. L. Kirkwood.
p. cm.
Includes bibliographical references and index
ISBN 0-19-513361-7
1. Aging. 2. Chance. 3. Individual differences.
4. Developmental biology—Mathematical models.
I. Kirkwood, T. B. L. II, Title.
[DNLM: 1. Aging—physiology. 2. Developmental Biology.
WT 104 F492c 2000]
QP86.F526 2000
612.6'7—dc21
DNLM/DLC
for Library of Congress   99-19723

9 8 7 6 5 4 3 2 1
Printed in the United States of America
on acid-free paper

Preface

We first met in January 1979 at a Gordon Research Conference on the Biology of Aging organized by George M. Martin. Over subsequent years an intermittent dialogue on stochastic issues in aging continued, mainly at various international conferences. We finally decided that it was time to turn this interaction into a book while attending Steve Stearn's 1997 conference Evolution in Health and Disease, in the delightful medieval Swiss village of Sion.

Each of us brought elements of this discussion that were developed independently (Finch, 1996, 1997; Kirkwood, 1996). But for each of us, the roots of our inquiry go back to our undergraduate years when an interest in chance variations was kindled. C.E.F. started to think about stochastic processes as an undergraduate in biophysics at Yale, particularly in 1960 during Ernest Pollard's marvelous course in thermodynamics and statistical mechanics. Later as a graduate student at Rockefeller University, he was intrigued by the individual differences in aging among inbred C57BL/6J mice, whose body weight and organ pathology he was following longitudinally. His inspiring adviser Alfred Mirsky also pointed out the importance of differences between identical twins. These experiences led directly to the longitudinal studies of reproductive aging and of mortality patterns that are central examples in our book.

T.B.L.K. entered biology via mathematics and biostatistics and was first inspired to think about stochastic processes in biology by a brilliant undergraduate course at Cambridge (England) taught by David Kendall. Subsequent graduate work in the department of biomathematics at Oxford brought contact with Maurice Bartlett, Michael Bulmer, and Roy Anderson. Interests in mechanisms and evolution of aging were triggered by collaboration with Robin Holliday and by inspirational discussions with John Maynard Smith. This led in the late 1970s to development of the "disposable soma theory" that provides a framework for the role of stochastic factors in development and aging.

In this book, we inquire about a little-explored subject: the sources of individual variations in postnatal development and in aging that cannot be attributed to genes or the external environment. Thus, many outcomes of aging may vary widely within an age group in ways that are not, according to these principles, predictable for any individual. Our goal here is to articulate issues about the role of chance in life history that we consider to be largely neglected by mainstream biomedical research. Of course, epidemiology and evolutionary biology recognize the role of chance explicitly. However, the role of intrinsic developmental variations has been relatively neglected in life history studies. We hope to expand the domain of theoretical and experimental approaches in biology generally with the agenda developed here to understand individual outcomes in aging.

Los Angeles, California   C.E.F.
Manchester, England   T.B.L.K.
January 1999

References

Finch CE (1996) Biological bases for plasticity during aging of individual life histories. In: *The Lifespan Development of Individuals: A Synthesis of Biological and Psychosocial Perspectives*, Nobel Symposium (ed. Magnusson D) pp 488–501, Cambridge University Press, Cambridge.

Finch CE (1997) Longevity: is everything under control? Non-genetic and non-environmental sources of variation. In *Longevity: To the Limits and Beyond* (ed. Robine J-M, Vaupel J, Jeune B, Allard M) pp 165–178, Springer-Verlag, Heidelburg.

Kirkwood TBL (1996) Human senescence. *BioEssays* 18: 1009–1016.

Acknowledgments

We are grateful for many colleagues who have contributed by various discussions of their published and unpublished data, and/or reading of various parts of the text: Richard Bergman (USC), John Breitner (Johns Hopkins U), Carl Cotman (UC Irvine), Kaare Christensen (U Odense), Vincent Cristofalo (Lankenau Medical Research, Philadelphia), Jonathan Cooke (NIMR London), Eric Davidson (CalTech), Malcolm Faddy (U Canterbury, New Zealand), Margie Gatz (USC), David Gaist (U Odense), Myron Goodman (USC), Roger Gosden (U Leeds), Lionel Harrison (U British Columbia), Tom Johnson (U Colorado), Marja Jylhä (U Tampere), Valter Longo (USC), Marion Meyer (Johns Hopkins), George Martin (U Washington), Jonathan Minden (U Pittsburgh), James Nelson (U Texas, San Antonio), Brenda Plassman (Duke), Chris Potten (Paterson Inst., Manchester), John Peters (USC), Michael Rose (UC Irvine), Ellen Rothenberg (Caltech), David Shook (U Virginia), Larry Squire (UC San Diego), Larry Swanson (USC), Fred vom Saal (U Missouri, Columbia), Martin Tenniswood (U Notre Dame), Rahul Warrior (USC), Michael Waterman (USC), Robert Williams (Vanderbilt U), John Tower (USC), and James Vaupel (MPI, Rostock).

We are particularly indebted for the major input by our loyal colleagues, who read most of, or all of, the full text: Jonathan Cooke, Eric

Davidson, Malcolm Faddy, Roger Gosden, George Martin, Ellen Rothenberg, John Tower, Rahul Warrior, and Robert Williams.

We are also grateful for the expert assistance of Linda Mitchell (USC) in assembling the bibliography and of Chris Anderson (USC) in graphics and drawing. At Oxford University Press, we were extremely fortunate for diligent and masterful editing by Lisa Stallings and for the enthusiasm and guidance of Kirk Jensen throughout the project.

Contents

CHANCE, DEVELOPMENT, AND AGING

1

Chance and Its Outcomes in Aging

Genetically identical human twins and highly inbred laboratory animals show wide individual differences in patterns of aging and in life spans. As we will discuss in detail, the heritability of life spans is modest in these examples. Moreover, many genetic risk factors for adult-onset diseases show widely varying ages at penetrance. These individual outcomes of aging pose a major puzzle because, particularly in inbred laboratory animals, these differences cannot be easily attributed to either variations in the adult environment or genetic differences. In trying to critically evaluate these puzzles, we were led to consider a little-recognized aspect of the biology of aging: chance events during prenatal development contribute to outcomes of aging. We will propose *intrinsic chance* as a third factor to the conventional two factor model, which attributes genes and the environment as the main determinants of life history.

In particular, we consider an explanation of *individual* patterns of aging that takes account of the extensive chance variations in cell numbers and connections, in cell fates during differentiation, and in physiological patterns that arise during development and that continue to unfold during adult life. This inquiry forced us to scrutinize the impact of many aspects of chance in biological systems. The information assembled drew us to a new perspective on aging. Many aspects of chance are well known within various subfields of biology, but could have a larger role than has been appreciated for their major, collective contribution to outcomes of aging and, more generally, to the life history.

This inquiry also led us to examine various limits to genomic authority in life history that arise through the actions of chance. In considering limits to genomic authority in this context, we are not challenging the supremacy of genes in living systems. Rather, we are scrutinizing how, as a consequence of evolutionary forces, outcomes of aging may not be controlled with the same rigor as is earlier development. Even so, we find evidence for considerable developmental noise that has significance to life history outcomes.

To set the context for this inquiry into life history variations, it is helpful to consider some general aspects of chance in biology. Chance is omnipresent in living systems, being at once the source both of creative novelty in evolution and of corruption and decay. Without random mistakes in the copying of genetic information, new life forms would not evolve. Yet the very kinds of mutations that may occasionally introduce a useful novelty in the germline also underlie many harmful genetic abnormalities. During the lifetime of an individual, mutations of somatic cells may contribute to the development of autoimmune diseases and cancers. The double-edged sword of DNA mutations is widely appreciated. What is less well recognized is the pervasive role of chance at other levels of organization: from cells, to organs, to individuals, and to entire populations. The combined action of chance variations can affect how the organism grows and develops, how long it lives, and how it eventually dies.

By examining the role of chance in development and aging, we have begun to develop the framework for a theory of how chance fluctuations in developmental processes can have major consequences for the life histories of individuals. We do not suggest that chance acts independently of genome and environment. The three major factors—chance, genome, and environment—interact at many levels throughout the life span. Nor do we suggest that chance leads *inevitably* to variable outcomes in development and aging. In some contexts, we shall show that chance fluctuations are very tightly constrained, so that the level of chance variations is almost negligible.

Actions of chance pose many intriguing puzzles. The nematode worm *Caenorhabditis elegans* is, at a cellular level, subservient to the strictest genetic determinism. Each wild-type individual has the same number (959) of somatic cells. The lineage of these cells, as the adult develops from the fertilized egg, is known precisely, including the identity of 131 cells that are programmed to die during development. This remarkable characteristic of fixed adult cell numbers is found in several phyla, whereas most other invertebrates and vertebrates have variable numbers of cells (Davidson, 1994). However, among a population of genetically identical worms raised in a constant laboratory environment, life span is highly variable: some individuals die at 10 days while others will survive

30 days or more (Fig. 1.1A). The broad temporal distribution of daily mortality extends over 20 days (Fig. 1.1B).

The source of the variations in *Caenorhabditis* aging is unknown. We will discuss the possibility that the variation arises from subtle random differences during development or in adult life that eventually affect

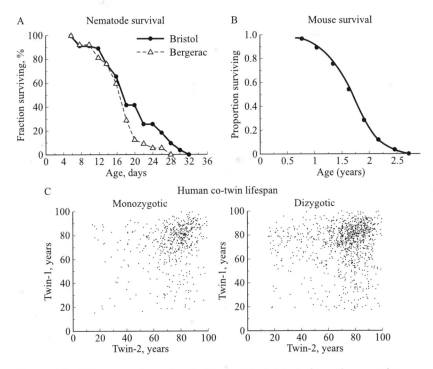

Figure 1.1. Life spans of genetically identical individuals from three species show extensive variations. *A*, Survival curves of self-fertilizing nematode (*Caenorhabditis elegans*), redrawn from Johnson and Wood (1982). Two laboratory strains: △, Bergerac; ●, Bristol. These populations were grown on liquid culture to minimize microenvironmental differences. Johnson and Wood (1982) comment, "Some variability in mean life-span was observed with apparently identical populations in experiments at different times. Because the life-spans of different stocks co-vary, we attribute these variations to still uncontrolled environmental effects" (p. 6604). Note that the range of these interexperiment variations in mean life span for each line is 1–3 days (5–15% of the mean). For comparison, the range of life spans of an inbred line within any experiment is 10–30 days (50% of the mean). *B*, Survival curve of inbred laboratory male mouse C57BL/6J strain. Redrawn from Finch and Pike (1996). *C*, Human identical (MZ) and fraternal (DZ) twins (females), showing differences in the life spans of co-twins. The correlations between life spans of co-twins is larger for MZ (0.20, 0.12–0.29, 95% CI) than DZ (0.06, 0.00–0.12, 95% CI); heritability—narrow sense, 1%; broad sense, 23% (Herskind et al., 1996). Data from Danish twin cohorts born 1870–1900, provided by Kaare Christensen of the Danish Twin Study (Herskind et al., 1996).

cellular functions and contribute to organismal death. An apparently similar intrinsic heterogeneity in aging patterns and life spans will be described for human twins and other genetically identical organisms. Among genetically identical individual humans and rodents, the patterns of aging in reproduction vary widely.

A reverse conundrum is found in periodical cicadas (*Magicada*), broods of which emerge from larval forms to mate as adults on either 13 or 17 year cycles (Finch, 1990, p. 75). How do different broods of cicadas know when their time is ripe? It is easy to imagine an evolutionary advantage of synchronous reproduction, but we know little of the physiological genetics that program these alternative cycles so precisely in the face of wide environmental fluctuations.

How these patterns of mortality, fertility, and longevity are shaped by natural selection is addressed by a well-developed evolutionary theory, which recognizes that the *force* of natural selection declines with age. As discussed further in section 1.7.3, the decline begins with the start of the reproduction and continues thereafter. The reason for the decline is that natural selection acts through gene variants that affect subsequent reproductive potential. Any gene effect manifested early in life is subject to maximum selection intensity, because reproduction lies in the future. However, a gene effect that shows up at a later age matters relatively less. Because the high levels of attrition in natural populations from external hazards allow relatively few to reach advanced ages, most of the reproduction is accomplished by young adults. Thus, a gene with late-arising adverse effects that cause cancer, diabetes, or Alzheimer disease would not be selected strongly against, because it is then too late to affect that part of reproduction which has taken place already, and because many who might have been affected will have already died before the gene difference can be expressed. This attenuation in the force of natural selection with age is thought to underlie the evolution of aging and argues powerfully against the idea that senescence itself is subject to strict genetic controls.

Nonetheless, each type of organism has a characteristic pattern of aging that is broadly generalized across genders, genotypes, and environments. For example, nematodes and fruitflies have no cancer or other abnormal growths during aging, in contrast to their common occurrence in humans and other mammals during aging. Thus, there is a basis for considering *canonical patterns of aging* (Finch, 1990), which are comparable in their species restriction to the canonical patterns of development upon which modern developmental biology has so depended. Tables 1.1 and 1.2 summarize the canonical patterns of aging in nematodes, fruitflies, and mammals for further reference.

We will also return in the coming chapters to the significance of the declining force of natural selection with increasing age. In particular, the contrast between the strong selection force on early life events and the weaker selection on later life events helps to reconcile the different

Table 1.1. Canonical age changes: major phyletic differences

	Oxidative damage to proteins	Tumors[a]	Neuron degeneration[a]	De novo oogenesis after maturation[c]	Midlife reproductive senescence in females	Vascular degeneration[d]
nematode	+	0	?	+	+	0
fruitfly	+	0	+	+	0	?
mammal	+	+	+	0	+	+

For general discussion and references, see Finch (1990, chapters, 2, 3, 7).

[a]The nematode *Caenorhabditis* and the fruitfly *Drosophila* have not shown tumors or other abnormal growths during adult aging, which is not surprising because of the absence of posthatching somatic cell division in *Caenorhabditis* and the very low levels of somatic cell proliferation in *Drosophila* gut and nervous system. The germ cell tumor mutants in *Drosophila* (Bae et al., 1994) and *Caenorhabditis* (Berry et al., 1997) cause sterility and are not normal phenotypes. The prevalence of tumors and other abnormal growths increases with age in all populations of mammals.

[b]*Drosophila* neurons show evidence of degeneration by the decreased numbers of fibers in the mushroom bodies (Technau, 1984) and loss of axonal branches in a mechanosensory neuron (Corfas and Dudai, 1991). Indrect evidence for neurodegeneration is the increased life span in transgenic fruitflies that overexpress human SOD1 in adult neurons (Parkes et al., 1998). In mammals, neuron degeneration takes several forms including atrophy of large neurons and decrease of neuron arbors (Finch, 1993). Neuron death, however, is generally sporadic in humans and laboratory rodents but is intensified in select brain regions during Alzheimer and Parkinson diseases (section 1.4.2).

[c]In *Caenorhabditis*, fecundity diminishes strikingly after 10 days and is followed by an extensive post-reproductive phase that may be 50% of their life span (Johnson, 1987). However, old hermaphrodites that have ceased oogenesis can be induced to reinitiate oocyte maturation by mating with a male (Ward and Carrel, 1979). Moreover, a multiply mated young hermaphrodite will produce up to fourfold the normal number of progeny (Kimble and Ward, 1988, p. 213). Although a major limiting factor in the duration of the fertile phase of adult life appears to be numbers of sperm, little is known about the proliferation of germ cells in aging worms (Samuel Ward, pers. commn.). In *Drosophila*, oogenesis continues during adult life without definitive age of cessation or the loss of fertility (e.g., Lin and Spradling, 1993, their Table 4) and older females lay eggs until just before death (Rose; 1991; Carey et al., 1998). In mammals, cessation of oogenesis before birth is a general characteristic, although there are suggestions that oogenesis continues in the adult ovary in dogs and certain prosimians (vom Saal et al., 1994; Finch and Sapolsky, 1999).

[d]*Caenorhabditis*, an acoelomate, has no vasculature. *Drosophila* has a heart and vascular system, but age changes have not been reported. Other insects show age-related vascular blockage, which could be due to blood cell aggregation and/or altered vessels (Finch, 1990, p. 65).

perspectives on life history within the fields of developmental biology and gerontology. Both of these disciplines are concerned with the timing of life history stages. But whereas developmental biology deals mainly with the regular and reproducible emergence of structures from the relatively simple fertilized egg, gerontology focuses on the deterioration of functions during adult life, resulting eventually in impaired homeostasis, disease, and death. Chance variations that arise during development may be relatively unimportant for their effects during early life, but can contribute substantially to variations in the outcomes that arise at older ages.

Puzzles we just considered, such as the constancy of cell number but variation of life span in *Caenorhabditis*, are in part resolved when we

Table 1.2. Canonical age changes in mammals

	Ovary[a] oocyte loss/fertility	Bone loss	Tumors and cancers by midlife	Benign prostate hypertrophy[b]	Atherosclerosis	Brain microglial/ astrocyte activation	Neuron atrophy[d]/D2 receptor loss[e]	Amyloid deposits in brain (Aβ)[e]
mouse/rat	complete/ midlife loss	M&F	M&F	+	+	+	+/+	0
dog	?/maintained	M&F	M&F	+	+	?	?	+
prosimian (*Microcebus*)	?/maintained	?	?	?	?	+	?	+
rhesus (*Macaca*)	complete/ midlife loss	M&F	M&F	+	+	+	+/+	+
human	complete/ midlife loss	M&F	M&F	+	+	+	+/+	+

Certain aging changes are shared by placental mammalian species across a 30-fold range of life spans, from mouse to human. Prominent age-related changes that are scheduled in proportion to species life spans include slow progressive increases in collagen oxidation in connective tissues and slow progressive loss of dopamine receptors in the brain (Finch, 1990). The shared subset of canonical aging changes in short- and long-lived species that descended from common ancestors 70–100 million years ago strongly implies genetic controls. Limited references are given, where not found in Finch (1990).

[a] The mammalian ovary is most generally characterized by its irreversible loss of irreplaceable oocytes from birth onward. Reproduction must cease when oocytes are depleted. The loss of oocytes is not necessarily complete in all species by the end of the life span. For example, female dogs and chimpanzees retain degrees of fertility until just before death in old age (Finch and Sapolsky, 1999; Finch, 1990). There are indications that de novo oogenesis continues in adults of a few species: dog (Anderson, 1970) and a different prosimian than cited above (Kumar, 1974, Sapolsky and Finch, 1999).

[b] Benign prostatic hypertrophy is common in aging humans (section 2.6.2), rhesus monkey (Baskerville et al., 1992; McEntee et al., 1996), and dogs (Juniewicz et al., 1994; Prins et al., 1996). BPH is reported in aging rats (Banarjee et al., 1998), but not aging mice (Maini et al., 1997); genotype influences are not defined within these species.

[c] Microglia and astrocytes show activated phenotypes during normal aging of laboratory rodents, rhesus monkey, and humans (Landfield et al., 1981; Morgan et al., 1999; Perry et al., 1993; Schipper and Stopa, 1995). Although glial activation is often associated with neurodegenerative diseases, its occurrence in healthy middle-aged rodents (Morgan et al., 1999; Schipper et al., 1981) suggests that this marker is of aging rather than of age-related disease. The attenuation of glial activation by diet restriction in aging rats (Morgan et al., 1999) indicates that local oxidative damage (a chance event) may be a factor: food restriction decreases the load of oxidized proteins and lipids, which can activate glia.

[d] Atrophy of large neurons is characteristic of aging and may be independent of sporadic neuron loss (Finch, 1993). At a biochemical level, there is gradual loss of the dopamine D2 receptor in the basal ganglia, which can be detected by middle age in humans and rodents (Morgan et al., 1987; Severson et al., 1982).

[e] Aβ amyloid deposits in the brain are found by middle age in all mammals so far examined that live more than 10 years (Finch and Sapolsky, 1999; Selkoe, 1994; Price et al., 1992). The amyloid deposits accumulated with "normal aging" are more diffuse than those found in Alzheimer disease, where there is also much greater neuron death. For neuron loss during aging, see Table 1.1. Laboratory rodents do not accumulate brain Aβ amyloid deposits, unless made transgenic with human Alzheimer mutations.

consider the interactions among chance, genes, and environment and how these interactions are modulated by age. Natural selection, in which environmental factors play a major part, influences the evolution of genes, some of which will limit the actions of chance. But the effects of chance on certain aspects of development matter more, and are therefore more tightly constrained, than chance effects on many outcomes of aging. There is no paradox in asserting that some outcomes of aging have their roots in actions of chance that first occur during development but that are relatively unimportant until later.

Chance is an essential ingredient of the chemical reactions that sustain life. Molecules moving through intracellular space are subject to all kinds of random encounters; some, such as collisions with free radicals, are highly destructive. The very interactions between molecules that influence key outcomes in cell differentiation and development are inescapably governed by chance. We wish to know at a fundamental, mechanistic level how chance affects the development of tissues, and how these fluctuations can have a long-term impact on viability and fertility throughout adult life. The recognition that chance is fundamental to subatomic mechanisms was a turning point in physical sciences in the early twentieth century. We anticipate that resolving the role of chance as it affects organisms at all stages in life history will profoundly influence biology in the next century.

Three terms—"chance," "random," and "stochastic"—are used often in our text. These words have overlapping meanings, but they are not completely equivalent. We use "chance" as the most generic term to describe the generating force that lies behind the variations that cannot (at least, not yet) be explained. "Random" describes outcomes that are essentially unpredictable. "Stochastic" is often used by statisticians and mathematicians to describe a process that generates a distribution of outcomes, which contain an element of chance or randomness. To the extent that random and stochastic differ in meaning, one may say that a process is stochastic if the outcome is a random variable whose distribution can in some way be mathematically specified. For example, the number of mutations arising in a population of cells on a given day is random, but the distribution of the daily numbers of mutations recorded across a period of several days has a particular statistical form, the Poisson distribution. The underlying mutational process is described as a stochastic process. A further term—"chaos"—is sometimes also encountered. Strictly speaking, a chaotic process need not involve the action of chance at all, and may even be fully deterministic. The hallmark of chaos is that the long-range outcome of a process cannot be predicted reliably because of extreme sensitivity to initial conditions, which cannot be determined exactly. The trajectory followed by a small particle lurching through random walks in solution according to Brownian motion is chaotic rather than random. Each collision is essentially deterministic (at least if we ignore quantum fluctuations) but their number is so large that they cannot all be predicted. For most practical purposes, it is hard to distinguish chaotic from truly random behaviors, and the difference—if it does exist—does not materially affect our conclusions. For an illuminating discussion of randomness in the context of computer-generated sequences, see Pincus and Singer (1998).

Finally, we acknowledge that, in some examples, that which is unexplained variation today may become explained tomorrow. The paradigm shift of physics early in the twentieth century was acceptance that quantum uncertainty, as enshrined in the famous Heisenberg Uncertainty Principle, is intrinsic to physical measurement and laws. The Uncertainty Principle derives from the wave–particle duality of the fundamental particles of matter, and reflects the fact that observation of a system necessarily perturbs it. For example, to measure the position of a particle in space involves interaction with a detector, which renders the subsequent motion of the particle unknowable beyond a minimum degree of uncertainty. A similar problem, but not of such an irreducible nature, arises in biology. We face major technical difficulties in observing the subcellular biochemical reactions of living systems, as they occur in vivo, without perturbing them so much that we cannot properly track their consequences. The difficulty is compounded if we wish to observe subcellular *variations* and their subsequent effects, simply because the amount of data needed is so great. No matter how well we understand how the genome specifies the organism, and the modulations by the external environment, we do not expect to predict *all* outcomes of aging, or for that matter of any other aspect of development with complete certainty.

This unpredictability is even true, as we shall see in Section 1.6.2, for the growth of bacterial populations because of intrinsic chance variations in the molecular mechanisms that control cell growth and division. Unlike the Uncertainty Principle in physics, however, it is theoretically possible to resolve biological variations with appropriate techniques and experimental designs. Our customary approach as biologists is to minimize "biological variations" in our experiments, through careful design and strict control of the experimental environment. We are less attuned to the possibility that the chance variations may themselves contain information that is valuable.

Readers should anticipate that the examples we discuss are not comprehensive. Moreover, despite many rigorous treatments in life history theory that consider species comparisons, there is no comparable treatment of the physiological and molecular factors that determine *individual* outcomes of life history, which requires much better understanding of the mechanisms at work. We are concerned throughout to present the uncertainties in the examples discussed and that we do not elaborate unnecessary hypotheses.

1.1. Main issues

Resolving the nature of individual differences is among the most profound, long-standing problems in biology: it is particularly difficult for complex long-lived organisms which are exposed to myriad environments at different stages of their life history. Beginning in the nineteenth century, long before the emergence of molecular biology, geneticists examined the inheritance of quantitative traits such as body size, growth rate, and fecundity. The classical model developed by Ronald Fisher, Sewall Wright, and J. B. S. Haldane in the first half of the twentieth century (Falconer and Mackay, 1996; Lynch and Walsh, 1998) resolves individual differences in quantitative traits in a population (the phenotypic variance, V_P) as two main terms, heredity (V_H) and environment (V_E):

$$V_P = V_H + V_E \tag{1.1}$$

Each of these can be resolved further (we ignore covariance terms at this point). V_E in human twins, for example, may be partitioned into shared and nonshared environmental components (Falconer and Mackay, 1996; Lynch and Walsh, 1998; Neale and Cardon, 1992). Nonetheless, despite the most exacting experimental approaches with almost complete elimination of genetic variation and the most strictly controlled environments (many examples follow), there remains an "intangible" variation, which has been recognized by several authorities:

> large amount of variability which is found to be due neither to heredity nor to tangible environmental conditions. (Wright, 1920, p. 328)

> Some of the intangible variation . . . may arise from "developmental" variation: variation, that is, which cannot be attributed to external circumstances, but is attributed, in ignorance of its exact nature, to "accidents" or "errors" of development as a general cause. Characters whose intangible variation is predominantly developmental are those connected with anatomical structure, . . . such as skeletal form or . . . bristle number in *Drosophila*. (Falconer and Mackay, 1996, p. 135)

We propose to introduce a third term, V_C, into equation 1.1 to account explicitly for chance variations in the development of genetically identical individuals (see Note, p. 75):

$$V_P = V_H + V_E + V_C \tag{1.2}$$

In the classical model (eq. 1.1), V_C is subsumed under V_E. Our discussions of chance variations during development include two major categories of chance events which occur from the beginnings of development and continue into adult life: those that arise during morphogenesis and differentiation and are part of developmental programs versus those that arise through random damage to DNA and other molecules throughout life. We will argue that the role of intrinsic chance variations arising during development (and, in some cases, continuing to unfold during adult life) require separate and explicit recognition.

We will look into the sources of chance variations during development that may be associated with individual outcomes of aging. In particular, we will examine evidence for individual variations in form and function that arise from chance events during development. By variations in "form," we mean anatomical variations in the number of cells, and also variations in cell–cell contacts, for example, the density of neuronal synapses. By variations in "function" we mean organ capacity, but also the individual set-points in regulation of hormones and metabolism, for example, the sensitivity of nervous system responses to hormones. By variations in "development" we mean prenatal (or prehatching) stages and postnatal stages up through sexual maturity. The developmental times range from hours to years, being even longer than a century in some plants (Finch, 1990, 1998). We will also consider the role of chance factors such as random damage to DNA and proteins that affect ongoing homeostatic functions in the adult. Even when morphogenesis itself is complete, chance events continue to affect the organism at a whole range of levels. Examples include adaptation of the immune system to novel antigens, imperfect replacement and renewal of damaged molecules and cells, and somatic mutations resulting in cancer. The prefertilization (prezygotic) phase may also influence adult characteristics across several generations (see section 2.7).

How might variations arise during these extended realms of development? The number of different environmental factors is vast and the number of possible interactions over time and space is infinite. The external environment of the developing individual includes obvious important factors subject to chance variations, such as nutrients, hormones, and toxins (chapter 5). We do not address the external environment at post-natal stages—our focus is on earlier development.

Even if external environmental factors are held constant, other variations arise *within* the developing individual. For example, the daughter cells from a single dividing cell may each become different types of cells; sometimes one cell dies and the other survives. Depending on the stage of development and the tissue, the differences in the fates of daughter cells are often observed to be determined with fixed but not necessarily equal probabilities. The role of the "cell-fate dice," loaded or not, can yield different numbers of cells to emerge during successive cell divisions in differentiation (chapters 2 and 3). Moreover, some of these chance-dependent events continue to generate molecular and cellular variations, processes that are poorly understood. There are powerful monitoring processes that constrain outcomes of differentiation, not only during development but also throughout the life span. A special source of variation in adult rodents arises through fetal interactions (section 4.1.2).

At a *molecular* level, other variations arise from the random walk, or Brownian motion, of molecules in solution. We will develop the hypothesis that Brownian motion–type fluctuations arise during diffusion of molecules that, in turn, cause variations during development in biosynthesis, gene activity, cell numbers, cell survival, and so on. These and other sources of chance variations that are not genetically programmed will modulate individual outcomes in aging.

To conceptualize these different levels of the environment, you may imagine a hypothetical thermometer sitting outside, which is warmed by the sun during the day and cools off at night. Gradual temperature changes in response to the external environment would be seen with an ordinary thermometer. Now let the tip of the thermometer shrink drastically—you would see increasing fluctuations in the temperature signal as the tip approached the size of molecules in the environment. These fluctuations would represent the random differences in the numbers and velocities of molecules hitting the tip, and correspond to the random walks of colloidal particles in classical Brownian motion. If this imaginary experiment were extended to a living cell, the shrinking probe would also encounter Brownian motion at many locations, such as in single membrane protein molecules (Sako and Kusumi, 1994) and in chromatin and other larger assemblies (Marshall et al., 1997). However, at any instant, few molecules within a cell are truly "free," that is, not associated by weak forces to other neighboring molecules. Diffusion within cells is constrained and guided by complex membrane systems and by scaffolding proteins (Nickerson et al., 1995; Mastick et al., 1998), for example, which confer specificity in kinase signaling pathways (Garrington and Johnson, 1999). Even so, the thermodynamics of diffusion predicts a constantly changing distribution of velocities, which will lead to varying local concentrations of molecules, within cells and in the extracellular spaces. Thus, intrinsic Brownian motion creates a noisy background for complex functions such as gene expression and cell migration, which, in turn, inevitably show some degree of intrinsic variation.

We will develop hypotheses about how certain individual differences in outcomes of aging may be attributable to certain variations in body form and functions that arise during development. In essence, chance variations in the internal and external environment, including the spatial distribution of crucial molecules, influence the number of cells produced and the characteristics of cell–cell interactions. These findings also confront us with a major challenge: the genetic library, no matter how precisely it is copied from cell to cell, still allows degrees of randomness as cells carry out their genetic instructions. Thus, all vertebrate animals, including identical twins, will be shown to vary in the numbers and organizations of cells in the brain and other body parts. At entry into the external world, individuals, even if genetically identical, have already begun different quantitative variations in form and function that must be considered in evaluating the impact of further random events, internal or external. The examples that led us to explore these possibilities may allow generalizations to other organ systems and organisms. In the examples discussed, we try to evaluate experimental (observer) errors.

We return briefly to the classical model from quantitative genetics (eq. 1.1), in which V_C is subsumed under the broad category of V_E. While there are strong experimental approaches to investigating V_H, it is harder to resolve the residual part of V_E that cannot be assigned to tangible environmental variations. In human genetics, a powerful approach to the developmental environment is analysis of twins separated at birth and of adopted children in the same family (Plomin, 1994; McGue and Bouchard, 1998). However, these approaches do not address the non-shared prenatal environment, of which certain components can be experimentally evaluated by analysis of the unfertilized egg and by embryo transfer (chapters 2 and 4). Even so, contributions to the adult phenotype from intrinsic developmental noise are hard to resolve. We give many examples of developmental variations that arise independently of the environment external to the organism.

Nonetheless, the external environment does contribute to developmental noise. Under some conditions, external stressors can increase the *fluctuating asymmetry* of bilateral traits (see sections 1.5.3, 4.3.1). It is fair to say that there is relative neglect of the contribution of variance to the phenotype due to chance variations during development (V_C). The conventional view which recognizes only V_H and V_E is confusing because it combines, inappropriately in our view, the contributions of intrinsic chance with "environmental" factors. Later in this chapter, we consider how chance variations in form and function influence outcomes of aging by setting the thresholds for the levels of age-related changes that can be compensated before dysfunction emerges.

1.2. Life span variations

Common species show wide differences in life spans under well-defined environments. Thus, mice live five times longer than their fleas, cats five times longer than mice, and humans five times longer than cats. The life spans of multicellular organisms differ nearly a millionfold (Finch, 1990, 1998). No one would doubt that genes directly determine these species differences in life spans, whether measured as life expectancy at maturity or the maximum longevity recorded. Yet genetic studies of identical human twins and of the laboratory mouse, fruitfly (*Drosophila melanogaster*), and nematode (*Caenorhabditis elegans*) agree that *within* each of these species, the majority of the differences among individual life spans can *not* be attributed to genetic variation.

Genetic estimates for the heritability of life span, defined as the proportion of variance in life spans that is inherited (V_H in eq. 1.1), range from 10% to 35% (Table 1.3). This calculation represents the "narrow sense" heritability, or the "additive genetic" component. The remainder of variance is divided into gene interactions (epistatic components) and environment, both shared and nonshared, in the case of human twins. A benchmark organism for genetic studies of life span is the soil nematode *Caenorhabditis elegans*, in which the heritability of life span was estimated from matings of self-fertilizing lines (Johnson and Wood, 1982). These studies and others represented in Table 1.3 used linear models to

Table 1.3. Heritability of lifespan

	Life span		
Species	Heritability[a]	C-VAR[b]	Mean
Nematode			15 days
within line	0%	34%	
between lines	34%	19% (16–24%)	
Fruitfly			40 days
within line	[<0.01%]		
between lines	6–9%	11%	
Mouse			27 months
within line	[<0.01%]	24% (19–71%)	
between lines	29%	16%	
Human twins	23–33%	MZ, 19%	75 years
		DZ, 29%	

[a]This table, adapted from Finch and Tanzi (1997), shows "narrow-sense" heritability that represents the additive component, except for the analysis of Danish twins (Herskind et al., 1996), in which the best fit was obtained with a model of dominant heritability and nonshared environment. These studies employed a linear model to partition the heritability of life spans into additive, dominant, and epistatic components. For details of calculations, see the following sources: nematode (*Caenorhabditis elegans*), Bristol hermaphrodites (Johnson and Wood, 1982; Brooks and Johnson, 1991); fruitfly (*Drosphila melanogaster*), 25 genotypes (Promislow et al., 1996); mouse (*Mus musculus*), 20 inbred lines (Gelman et al., 1988); human twins from Denmark (Herskind et al., 1996) or Sweden (Ljungquist et al., 1998).

[b]The coefficient of variation (C-VAR) is a dimensionless number that represents the standard deviation (SD) as a percentage of (normalized to) the mean (X), in this case the life span, calculated as $(SD/X) \times 100$.

[c]Further post hoc analysis of the mice from the study of Gelman et al. (1988) showed that these 20 recombinant inbred strains include variant sequences on chromosome 11 that modified the range of their life spans by about 40% (difference in age from the longest- to the shortest lived individual in each strain) (de Haan et al., 1998). These strong effects were observed across the full range of mean life span among the strains. Ongoing mapping studies may resolve the gene(s) in a quantative trait locus (QTL) (Falconer and Mackay, 1996; Lynch and Walsh, 1998) which generates variability in the complex pathophysiology underlying adult mortality rates.

estimate genetic and environmental variance. Estimates of broad- and narrow-sense heritability of the life span averaged 35% (range of 19–45%) for all species.

A caveat in interpreting these estimates of heritability is that even extensive inbreeding does not guarantee genetic identity. During inbreeding of sexually reproducing species, brother–sister mating for consecutive generations produces an asymptotic approach to isogenicity, for example, by 100 generations, with a limit determined by the mutation rate. For *Caenorhabditis*, the standard calculation predicts that 20 generations of self-fertilization will reduce genetic variation to 1 in 20,000 genes (Johnson and Wood, 1982). The main source of genetic variations expected is spontaneous mutations, which are rare for chromosomal genes, $< 10^{-5}$ per generation. However, mitochondrial DNA replicates more frequently and has a higher intrinsic frequency of mutation. (For other sources of genetic changes that can arise during inbreeding, see table 1.4, note g.) Thus, comparisons of the heritability of life span from self-fertilizing worms with that of inbred lines of flies or mice must accept some uncertainty in residual genetic variation. Suffice it to say, few geneticists would claim that the usual characterizations of inbreeding by graft tolerance, or by immunoepitopes, isozymes, or other polymorphisms which sample but few genes could guarantee absolute isogenicity.

To assist comparisons of the variations in life spans among species with major differences in life span, we show the coefficient of variation (C-VAR), which is simply the standard deviation as a percentage of the mean (Table 1.3, note b). The C-VAR for life spans is at least as great for certain inbred laboratory species (nematodes, flies, mice) under controlled conditions as that of human identical (monozygous, MZ) twins who are exposed to a far greater range of environmental fluctuations during life spans that are orders of magnitude longer, for example, C-VAR of the nematode Caenorhabditis life span (34%) versus human MZ twins (19%). The variance in life spans may also be under genetic influence within a species, as indicated for recombinant inbred mice (de Haan et al., 1998) (Table 1.3, footnote c).

The 34% C-VAR for life span in Caenorhabditis represents the several fold range of individual adult life spans in the same culture dish (Fig. 1.1A). A similar range of life spans is shown in the survival curves of longer lived mutant strains, which sometimes overlap with the distribution of life spans in various baseline strains (e.g., Apfeld and Kenyon, 1998; Gems et al., 1998; Ishii et al., 1998; Lakowski and Hekimi, 1998; Tissenbaum and Ruvkun, 1998). Besides these variations in life span, individual worms differ widely in the schedule of age changes in reproduction and adult swimming movements (Bolanowski et al., 1981; Johnson, 1987; Brooks and Johnson, 1991; Duhon and Johnson, 1995). Even young adults varied individually in their numbers of progeny (eggs that produced viable adults), with C-VARs of 20–30% (Brooks and Johnson, 1991). The day-to-day variations in fertility were generally correlated in individual worms. However, in another study, the swimming movement rates of individual worms showed "no consistency for the day-by-day correlations. Thus, although swimming rate is sometimes correlated with later movement rate (during aging), we cannot reliably predict an animal's movement based on movement on a previous day . . . " (Duhon and Johnson, 1995). We estimated a C-VAR of 70% for swimming movements from their graphed data. Possible causes of these extensive individual variations within an inbred strain of C. elegans are rarely discussed.

If these variations are not genetic, then they should be environmental, according to the conventional model (eq. 1.1). However, the population shown in Figure 1.1A was grown in liquid–phase culture dishes with live bacteria of a defined strain as a food source (Johnson and Wood, 1982). Liquid cultures that were constantly mixed by shaking gave the same range of life spans (Vaupel et al., 1998, their Fig. 1; Tom Johnson, pers. comm.). We can therefore confidently rule out that different life spans of *Caenorhabditis* might arise from local inhomogeneities in the availability of bacteria as food, or in amounts of toxic waste products.

If not genetic and if not environmental, what, then, could cause these major differences in aging of individual worms? The informed reader might first consider random damage to somatic cells, such as oxidative damage to proteins from free radicals (Ishii et al., 1998) and mitochondrial DNA mutations (Melov et al., 1994). These observations are based on whole worm extracts, as are most studies on aging invertebrates, and do not yet inform us about relationships of these molecular lesions to particular cell dysfunctions, which must differ among individual worms. Ongoing work may give such insights. The extensive individual differences in functions of individual young adult worms, however, are not easy to attribute to molecular aging processes, unless damage has already accrued during development.

We also ask if individual differences in worm aging might be associated with subtle individual variations in subcellular anatomy and cell physiology that arose from developmental noise, for example, chance variations in cell–cell contact that influence cell function. This suggestion may seem gratuitous to readers who know that *Caenorhabditis* has remarkably constant numbers of cells (see introductory remarks). However, fine-structure analysis shows that, while cell numbers do not vary, the locations of cell bodies and the types of cell–cell contacts do vary among individuals (discussed further in section 1.4.4). Moreover, young adult individual worms have significantly different levels of heat shock gene expression, as evaluated by reporter constructs with the *hsp-16* promoter (Tom Johnson, in prep). No one has examined how such subtle variations influence organ function. It is now feasible to do longitudinal studies using the new high-resolution optical techniques to look for relationships between subcellular architecture, gene expression, and aging in these transparent worms.

Extensive individual variations in worm movement frequency are noted above (Duhon and Johnson, 1995). Because the pharyngeal pumping rate is linearly correlated with food ingestion (Klass, 1983), it would be interesting to know if individual differences in swimming among worms were correlated with rates of pumping and of bacterial ingestion. These questions are already being considered in relation to the *eat* mutants that have impaired pharyngeal pumping. Lakowski and Hekimi (1998) showed that *eat* mutants in 7 of 10 genes had longer life spans, consistent with the extension of life span by food restriction in *Caenorhabditis*. We suggest further that variations of pumping among individuals in nonmutant strains might be equivalent to moderate degrees of food restriction. This can be tested directly. An alternative hypothesis involves individual variations in oxidative damage during aging due to monoamine neurotransmitters, which can generate free radicals (e.g., Levay and Bodell, 1997; Jenner, 1998). Individuals could vary in the density of monoaminergic neuron synapses, in levels of expression of the monoamine transporter (Duerr et al., 1999), or in antioxidant defenses. Although these suggestions are speculative, there is much more evidence for anatomical variations during development that have an impact on aging in reproductive organs of mammals (see below and chapter 2).

MZ co-twins also have widely differing individual life spans (Fig. 1.1C). For those twins who survive to 60 years, the correlation of life spans between MZ twin pairs is only slightly greater than that for di-zygotic (DZ) twins derived from two separately fertilized eggs, who are no more closely related than other brothers or sisters. From twin pair correlations, the heritability of life span is about 25% (Table 1.3). Two studies of MZ and DZ twins that survived to very old ages agreed that environmental factors are more important than heredity in life expec-tancy and many specific aspects of function. In Swedish MZ twins who were reared apart (130 pairs) versus together (1734 pairs), the heritability of life spans in MZ twins was <35% (Ljungquist et al., 1998). Deaths before 35 years were excluded; life spans ranged up to 90 years. About 50% of co-twin deaths occurred within 2 years, with a range of more than 10 years. Despite this wide range of times of separation, the only indi-cated genetic effects in MZ twins reared apart were in females. Males show no genetic influence on life span. A larger sample of Danish twins also showed low heritability, but without consideration of rearing together or apart (Herskind et al., 1996). Both twin studies concluded that most (65%) differences between twin life spans were due to non-shared environment.

The "nonshared environment," or individual experience, is usually attributed to variations in the adult environment. Nonetheless, the non-shared environment of the postnatal years and the nonshared environ-ment of the maternal uterus during gestation still allow vast possibilities for different individual exposure to external influences, such as arise prenatally from fetal blood supply (mono- vs. dichorionic MZ twins) and the duration of labor (birth order), and postnatally from social inter-actions. Variations in the uterine environment are discussed in chapter 5. Besides these frequently discussed sources of variation, we will argue for the importance of chance events within the embryo that lead to different cell numbers and cell–cell contacts.

Last in this discussion, we emphasize that the estimates of heritability for life span are in the same range for other life history traits analyzed by population biologists: fecundity and the duration of the reproductive schedule (Price and Schluter, 1991; Roff and Mousseau, 1987; Lynch and Walsh, 1998, p. 175). The term "life history trait" refers to a major target of natural selection. For example, selection pressure from predators of adults might select for genes favoring early onset of repro-duction, thereby reducing the exposure of adults (Austad, 1993, 1997). The reproductive schedule (e.g., age of onset; frequency of reproduction) and associated reproductive variables (e.g., egg and milk production) are

common life history traits that respond rapidly to experimental selection (Rose, 1991; Charlesworth, 1994; Stearns, 1992). Under such conditions, the statistical life span is an indirect outcome of selection for fecundity and the age of maturation. The heritability of a trait is an estimate of the genetic diversity in the population that codes for individual differences. When an optimal genotype has been selected, then the genetic variations in the population often decrease for the selected traits, which in turn reduces their heritability as observed in breeding studies. These processes are represented in Fisher's fundamental theorem (Rose, 1991; Charlesworth, 1994; Lynch and Walsh, 1998; Falconer and Mackay, 1996).

1.3. Variation in reproductive aging

Aging of reproductive functions varies widely among individuals in all organisms for which data are available. We emphasize rodents in this discussion because the depth of experimental studies allows us to develop key concepts. Reproductive aging in laboratory mice and rats resembles that of humans, with a loss of fertility cycles during midlife, which for these species can be considered a canonical change of aging (Table 1.2). The loss of ovarian oocytes during aging is a major cause of age-related infertility, as discussed in section 1.4.1.

In more than 10 genotypes of inbred rodents, wide individual differences are observed in the age when fertility (estrous) cycles become less regular (vom Saal et al., 1994; Lerner et al., 1988). In general, fertility cycles eventually cease by about 6–12 months before the end of the individual life span. Figure 1.2A represents the daily fluctuations of vaginal cell types, which can be conveniently and noninvasively monitored in the loose vaginal cells obtained from lavages (washing out the vaginal contents with a drop of saline). The vaginal cytology is closely coupled to the regular fluctuations in blood levels of estradiol and progesterone that drive fertility cycles. Because age changes typically emerge during a period of 3–12 months (up to 35% of the life span), longitudinal studies are needed to document the individual changes. The more common, cross-sectional study design, which compares different age groups during several weeks of 5 days per week observations, will seriously underreport the range of individual variations.

In longitudinal studies from the laboratory of C.E.F., females from the same birth cohort were followed via daily vaginal lavages during their reproductive life spans without contact with males (Nelson et al., 1981, 1982; Gosden et al., 1983; Mobbs et al., 1984). My collaborators and I were greatly puzzled then (1975–1983) by the wide individual variations in this highly inbred C57BL/6J strain. This same genotype shows a similarly wide range of individual life spans (Fig. 1.1) and has been inbred by brother–sister mating for > 100 generations since 1936 at the Jackson Laboratory (Table 1.4, note a). Each mouse had a characteristic pattern

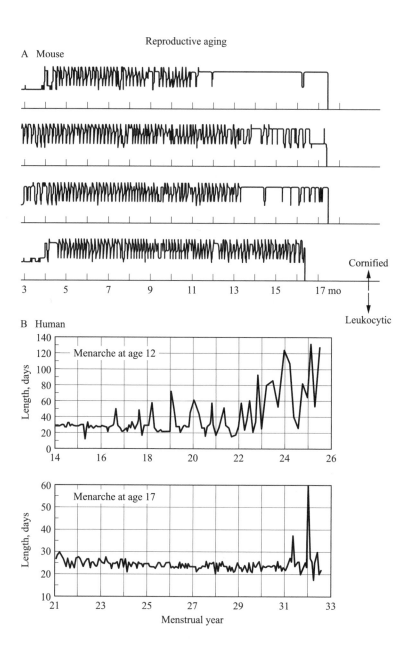

Figure 1.2. Variations in reproductive aging that cannot be attributed to
genetic variations. *A*, Individual longitudinal profiles of estrous (fertility) cycles
in C57BL/6J female mice. Inbred mice have different frequency and regularity
of estrous cycles as young adults and different transitions to acyclicity during
aging. The vertical scale represents the percentage of leukocytes and cornified
(keratinized) epithelial cells in daily records of vaginal lavages that contain
loose cells exfoliated from the internal surface of the vagina. Daily changes in
(*continued*)

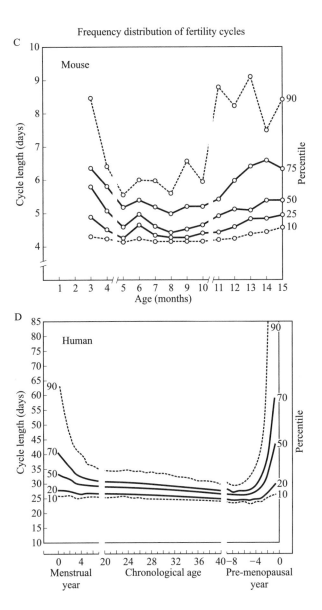

Frequency distribution of fertility cycles

Figure 1.2 (continued)
the vaginal cytology are driven by changes in ovarian secretions of estradiol
and progesterone. There are many individual variations in the transitions to
acyclicity: varying degrees of cycle lengthening; persistent estrus; persistent
diestrus, which can include pseudopregnancy; and anestrus, which is associated
with the cessation of ovarian steroid production and the exhaustion of oocytes
and follicles (vom Saal et al., 1994; Gosden, 1985). Redrawn from Finch et al.
(1980). *B*, Human female, individual longitudinal profiles of menstrual cycles at
the approach to menopause, from self-reports of alumnae of the University of
(continued)

21

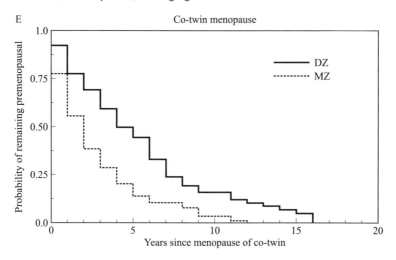

Figure 1.2 (continued)
Minnesota. The degree of cycle length variability before the onset of
menopause is an individual characteristic, as is the transition to menopause.
Redrawn from Treloar et al. (1967). *C*, Fertility cycle length frequency classes
of C57BL/6J female mice. Redrawn from Nelson et al. (1982). *D*, Human
female fertility cycle length frequency classes (University of Minnesota
alumnae); redrawn from Treloar et al. (1967). *E*, Differences between co-twins
in the age at menopause in the United Kingdom. Redrawn from Snieder et al.
(1998).

of cycle change during aging. Some ceased cycling by 9 months, whereas
cycles persisted in others to 18 months (Fig. 1.2A). The transitions to
acyclicity could be abrupt or gradual, and could proceed through several
alternative endocrine states (constant estrous, repetitive pseudopreg-
nancy/persistent diestrus, or anestrus), which are characterized by different ovarian output of estradiol and progesterone. These stages are
preceded, moreover, by prior individual variations in the frequency and
regularity of cycles from maturity onward. The variability among individual cycle lengths increases strikingly during aging (Fig. 1.2B). Moreover,
the strength of lordosis behavior during aging varies independently of the
ovarian status (LeFevre and McClintock, 1988). These individual diver-
gences suggest multiple locations of aging changes in the neuroendocrine
system (Finch et al., 1984; Wise et al., 1996). There is no clear relationship in laboratory rodents between the patterns of reproductive aging and
the life span.

    Human menstrual cycle changes before menopause also vary widely
among individuals. Figure 1.2B shows individual records (self-reports)
from Alan Treloar's landmark study of University of Minnesota
alumnae, 1934–1960 (Treloar et al., 1967). The variability in individual
cycle lengths also increases strikingly during aging, just as in rodent cycle

aging (Fig. 1.2D). Longitudinal studies of twins in the United Kingdom and in Australia show stronger correlations between ages at menopause for MZ than DZ twins; estimates of heritability (broad sense) for age at menopause range from 31% to 63% (Do et al., 1987; Snieder et al., 1998; Treloar et al., 1998a,b [this report is by Susan Treloar, grand-niece of Alan Treloar]; Do et al., in press). Nonetheless, in about 20% of MZ co-twins, the age at menopause differed by 5 or more years (Fig. 1.2E; Snieder et al., 1998). This extensive divergence in age at menopause in a subgroup of MZ twins is consistent with the wide range of ovarian oocytes in inbred mice as discussed below. There is no correlation between the ages of menarche and menopause among identical co-twins (Snieder et al., 1998; Kaare Christensen et al., unpublished).

These observations suggest that there is a strong nongenetic source of individual variations in the reproductive aging of female mammals. In laboratory rodents, environmental variations can be much more controlled than in humans. However, even if rodents are maintained one per cage to eliminate heterogeneity from social rankings, solitary housing is a stress by itself, that might be further individuated by chance early developmental effects on adult stress responses (sections 4.1.2 and 4.3). Moreover, environmental variations are not completely eliminated in conventional facilities, because of sounds and odors from nearby cages.

The confounding effects from environmental variations in laboratory rodents, however, may be judged as minor compared to the huge individual variations of cell numbers in the ovaries and in the neuroendocrine centers of the brain. There is a threefold range in the numbers of ovarian oocytes present at birth among individuals in inbred mice as well as in unrelated humans (Fig. 1.3), which are described further in section 2.2.2. We will discuss an experiment showing that young mice with fewer oocytes in their ovaries become acyclic at earlier ages, from which we argue that much of the variation in oocyte numbers arises from random features in germ cell differentiation and migration during the ovary development. Another source of variations in adult reproductive functions can be traced to the brain. In particular, reproductive controls that are seated in the hypothalamus differ within a single litter of inbred mice. These remarkable individual differences, in turn, can be traced to interactions between neighboring fetuses during pregnancy that alter fetal hormone levels during the differentiation of the hypothalamus (sections 2.6.1 and 4.1). Thus, the fetal interactions are a factor in the loss of fertility during rodent aging.

1.4. Variable numbers of irreplaceable cells

The degree of constancy in the numbers of cells in adult tissues varies widely among species. *Caenorhabditis* and certain other nematodes are at one extreme, with a constant number of cells characteristic of each species and no somatic cell replacement. *Drosphila* and some other insects have

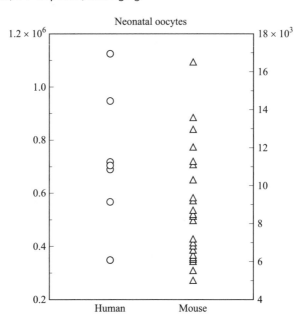

Figure 1.3. Variations in numbers of ovarian oocytes in neonates. *Left*, Human infants (Swedish, consanguinity unknown) most of whom died from asphyxia during labor. Follicles were counted every 200th section (20 μm), with a counting error of 8%, which was smaller than the threefold range in follicle number. "The number of primordial follicles seems to vary greatly in newborn infants" (Block, 1953). Drawn from tabular data of Block (1953). *Right*, Mouse (RIII strain). Redrawn from graphed data of Jones and Krohn (1961a).

limited but detectable cell proliferation in some adult tissues (Table 1.1). Mammalian tissues contain different proportions of cells that are continually replaced in adults but, of great importance to outcomes of aging, crucial cells that are never replaced. At the far extreme are sponges (coelenterates), which show little if any limitation in cell replacement, and whose cell numbers presumably differ widely among individuals, because of the indeterminate growth pattern and because of the capacity for vegetative reproduction by budding and fission.

The cells that produce sperm and eggs are called the "germ line cells," to distinguish them from the "somatic cells" of the rest of the organism. In mammals, oocyte numbers are limited because the primordial germ cells that give rise to oocytes are not replaced by new cells after birth (section 2.3). In contrast, the sperm-forming germ cells in the testes continue to produce new sperm throughout life. The initial stock of oocytes is progressively depleted by cell death in a process of apparently random loss with exponential kinetics, similar to radioactive decay. The loss of oocytes is progressive throughout life and eventually limits the duration of reproduction, as discussed in section 2.2.

In the case of the brain, the loss of neurons is both more sporadic and less extensive than the loss of oocytes in the ovary during normal aging, but can be severe as a consequence of diseases such as Alzheimer disease, Parkinson disease, or stroke. The adult brains of humans and other mammals have significant numbers of neurons that can proliferate (Eriksson et al., 1998; Gage et al., 1998). Variations in these endogenous neuron progenitor cells (the extent of individual differences is not clear) could be a factor in the onset or course of Alzheimer disease and other neurodegenerative conditions (see section 1.4.4).

Some of the shortest lived cells are found in the epithelial lining of the gut, where cells have life spans of a few days before they are shed into the gut. This high rate of cell turnover in the gut is sustained throughout life, yet the stem cells in each crypt of the intestinal wall, which are their source, are few in number and vulnerable to random loss (section 3.2.3).

In each organ, the extent of cell replaceability is established during development, through the activation of particular sets of genes. The molecular mechanisms that form the architecture of the body depend ultimately on selective activation of different sets of genes. Typically, many more cells are formed during development than survive to adult stages. In many examples, there is a great deal of variation in the numbers of cells that survive in adult organs. In the ovary and nervous system, we find different ranges of variations, which moreover allow different numbers of cells to form in bilateralized organs on the left and right sides of an individual. The examples discussed could be greatly extended. Our purpose at this early stage of our inquiry is not to be comprehensive, but rather to illustrate the kind of data available for scrutiny of individual variations.

1.4.1. Ovarian oocytes

The ovary at birth shows greater than threefold individual variation in the numbers of oocytes and primary follicles in humans and mice (Fig. 1.3). The possibility that these remarkable variations are not due to genetic variations is indicated by the same degree of variation in inbred mice, as first documented in the classical studies of Jones and Krohn (1961a,b). The individuals diverge further with advancing age, as the oocytes and follicles become depleted. These phenomena are discussed at greater length in section 2.2. The C-VAR for oocyte numbers of neonates or young adults is threefold larger than that for retinal ganglion cells in several strains (Table 1.4). The human data represent apparently unrelated individuals, obtained from ovaries of term stillbirths in Sweden (Block, 1953). We note that this and other studies of human ovaries used pathological material (see Fig. 1.3 legend). Although there are similar variations of oocyte numbers in healthy inbred mice, it is always necessary to consider possible confounds from morbid processes that are inherent to the study of human postmortem specimens.

Table 1.4. Cell numbers (×10³) in ovary and retinal ganglion of inbred mice

Strain[a]	Neonate oocytes[b]		Adult oocytes[c]		Strain	Retinal ganglion cells[d]		Brain weight (mg)[e]	
	Mean ± SD (range)	C-VAR	Mean ± SD (range)	C-VAR		Mean ± SD (range)	C-VAR	Mean ± SD (range)	C-VAR
A	9.1 ± 2.8 (4.4–16.5)	37%	3.6 ± 0.76 (2.7–4.5)	21%	A/J	50.5 ± 3.5 (44.9–58.0)	6.9%	411 ± 20.9 (381–464)	5.1%
CBA	7.2 ± 2.7 (2.5–11.5)	37%	2.6 ± 0.8 (1.5–4.2)	31%	C57BL/6J	54.6 ± 3.9[e,f,g] (47.1–60.7)	7.1%	498 ± 21.1 (452–542)	4.2
RIII	7.8 ± 3.3 (3.9–17.6)	43%	4.4 ± 1.1 (2.9–6.8)	25%	CBA/CaJ	56.0 ± 2.7 (47.7–62.4)	4.8%	438 ± 20.9 (401–478)	4.8
					DBA/2J	63.5 ± 4.2 (57.6–69.8)	6.6%	419 ± 27 (363–472)	6.5

C-VAR is coefficient of variation; see Table 1.3 b.

[a]Oocyte data were obtained from strains of mice maintained in the United Kingdom and derived >70 years ago. Lines maintained at the Jackson Laboratory may be separated by decades from those with the same symbol maintained in other independent colonies; standards for strain maintenance vary widely, for example, the NMRI strain used for analysis of synaptic spacing (Section 1.4.3) may be random- or pen-bred in some European colonies (Festing, 1996). Retinal data were obtained with mice of strains from the Jackson Laboratory (see note f). During the >200 generations of brother–sister inbreeding of these separate lines, it is likely that each line acquired variant genes not present in the original strains (see note g).

[b]Neonatal ovary (0–30 days). Numbers were extracted by Gul Seckin (U.S.C.) from graphed data of Faddy et al. (1983), using a window (boundary values) of 0–30 days. The same values (±10%) were obtained with a bigger window (0–50 days) and also with data from Jones and Krohn (1961a), who studied earlier lines of these same strains, also in the United Kingdom.

[c]Young adult ovary (50–100 days). Numbers were extracted by Gul Seckin from graphed data of Faddy et al. (1983), using a window of 50–100 days.

[d]Retinal ganglion of young adult mice (average age, 85 days; range, 21–329 days), cited directly from numeric data and statistics as calculated by Williams et al. (1996). There is no change of retinal ganglion cell numbers with age (Robert Williams, pers. comm.). The full range of values for individual mice with recounts is graphed in Figure 1.4A.

[e]Brain weight data as listed on the World Wide Web at http://mickey.utmem.edu:777/FMPro.

[f]Genetic variation affecting neuron number can arise rapidly even in a long-established strain. Williams et al. (1996) observed 20% differences in the mean numbers of retinal ganglion cells between two groups of C57BL/6J mice obtained from the Jackson Laboratory (Bar Harbor, Maine), which provides standardized inbred mice to researchers around the world. In contrast to the normative 55,000 retinal ganglia neurons that have been repeatedly observed in mice from several breeding colonies at the Jackson Laboratory, mice obtained from another location at the Jackson Laboratory (Annex I production colony) had 66,000 average neurons. Variant genes may have been introduced by rogue mice (no colony is ever secure from invaders for long). Alternatively, a mutation may have arisen and become fixed without being recognized. This may occur because genetic surveillance can only evaluate a small

fraction of the > 50,000 genes, such as those for the Mhc antigens or coat color. Besides point mutations, mice also are notorious for germline transmission of retroviruses, which can be unstable and, by position effects, may influence neighboring gene activities (see note g). For other examples of strain drift, see Finch (1990, pp. 318–319).

[g]The achievement of isogenicity, while asymptotically approaching 100% (Klein, 1976) is limited by the frequency of recombination and mutation, which varies among chromosomal regions. Chromosomal translocations also can modify gene expression through position effects. Retrovirus-like mobile genetic elements are known for rapid changes among closely related mouse strains (e.g., Fredholm et al., 1991; Weiss et al., 1989), and may also transport contiguous DNA sequences. Retrotransposable integration is proposed as a general mechanism of mutation, because of examples from cancer and other human genetic diseases (Miki, 1998). The insertion of transposable elements can also cause position (insertion) effects. Position effects can introduce confounds in experimental manipulation of gene dose, for example, as used to test the effects of antioxidant enzymes or protein synthesis factors on the life span of fruit flies (Kaiser et al., 1997). Even if future technology should enable efficient total sequencing of laboratory stock genomes, position effects would still be hard to predict. Some DNA sequences are unstable and prone to expand or contract within a few generations, for example, telomeric DNA repeats (Kipling, 1995; Kipling et al., 1996) and trinucleotide repeats that encode polyglutamine that can accumulate insidiously (section 5.1). Moreover, there is evidence that telomeric DNA may diverge rapidly within identical genotypes, as observed in somatic cell DNA from skin plugs of inbred mice (Kipling et al., 1996) and in lymphocyte DNA between identical twins (Slagboom et al., 1994). Genetic transmission of these individual differences remains to be shown. Last, mitochondrial DNA replication also is more error-prone than that of chromosomal replication, which can lead to heteroplasmy. These latter sporadic sources of variation will also arise in clonally reproducing organisms.

Genetic influences on menopause were discussed in section 1.3. Genetic influences on oocyte numbers are indicated in the rare familial syndromes of early menopause (before 40 years; Torgerson et al., 1997; Veneman et al., 1991; Cohen and Speroff, 1991). Inbred strains of mice also differ in the numbers of oocytes present at birth; for example, the CBA strain has a smaller endowment of oocytes (Table 1.4) and loses fertility 3–6 months before most other genotypes (Jones and Krohn, 1961a). The genes that determine oocyte numbers are not known, although several mutations modify oocyte numbers; for example, caspase-2–deficient mice have excess oocytes because of deficient cell death mechanisms (Bergeron et al., 1998), whereas mutations in the zinc-finger protein gene *Zfx* greatly decrease oocyte numbers (Luoh et al., 1997). There may be gene variants in human populations that modify oocyte numbers. We note the definition that common variants, existing in $> 1\%$ of a population, are referred to as "gene polymorphisms" to distinguish them from rarer "mutations," which arise sporadically.

1.4.2. Neuron numbers

Neuron numbers are more difficult to estimate precisely than ovarian oocytes, particularly in the closely packed neuron layers that are characteristic of the cerebral cortex and hippocampus. Counting artifacts are encountered less in the mammalian ovary, because each oocyte is clearly demarcated from its neighbors by one or more layers of enveloping follicle cells. For this discussion, we emphasize recent reports in which neuron numbers were estimated by stereological methods ("optical dissector"), which use thick sections and section depth calibration to minimize the double counting of cells (Williams and Rakic, 1988; West, 1991, 1999; Rasmussen et al., 1996; Simic et al., 1997).

A benchmark for precise estimation of neuron variations within a genotype is the retinal ganglion cell layer, in which neurons are arranged on a single layer, with axons that are separated by myelin sheaths and therefore easy to distinguish. The coefficient of error (standard error as a percentage of the sample mean) as determined by replicate counting is 2.5% for adults (Williams et al., 1996; Strom and Williams, 1998). The test–retest coefficient representing observational error is 0.83; thus, about 65% of the technical variance is observer error (Williams et al., 1996). The distribution was symmetric and near Gaussian.

Relative to the ovarian oocyte counts, there are modest individual differences in the numbers of retinal ganglion neurons in 10 inbred mouse strains (Williams et al., 1996; Strom and Williams, 1998; Table 1.4). The individual variations of retinal ganglion cell numbers within these strains ranged from 15% to 30%. Table 1.4 also shows brain weight variations for each of these strains, for which the *C-VAR* approximates those of retinal ganglion numbers. Figure 1.4 shows replicate measurements on the same mouse (five mice total; left) and the distribution of

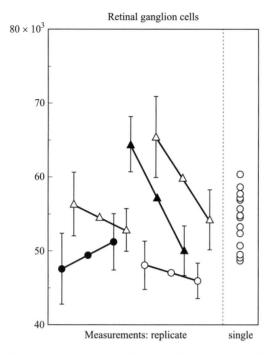

Figure 1.4. Individual variations of retinal ganglion neuron numbers in C57BL/6J mice. The number of cells is estimated from samplings of a thin cross section of the optic nerve, which contains axons exclusively derived from the retinal ganglia of one eye. The *replicate measurements* were made by different personnel on an adjacent section. Sampling was from the same retina in five mice, with means ±SEM, which represent two to four recounts by the same operator; the mean values of these independent estimates are connected by a line, with the average as a single point midway. For comparison, we also plot the *single measurements*, which span the same range. Data from Table 1 of Williams et al. (1996). These data are also shown in Table 1.3.

individual mean values (right). When experimental counting errors are accounted for, the *C-VAR* for nongenetic variations of neuron number among individual adult mice was calculated at 3.6% ± 0.4% (Williams et al., 1996), which is not inconsistent with the wider range of individual values shown in Figure 1.4. The intrinsic variation in neuron numbers within a strain was smaller than the mean differences between inbred strains, which ranged up to 20% (see examples in Table 1.4). Although several of these strains are closely related to those used to analyze oocyte numbers more than 30 years ago, they may not be identical genotypes (Table 1.4, note a). A genetic locus for strain differences in ganglion cells was mapped to chromosome 11, *Nnc1* (neuron number control 1), in which a single allele causes a difference of 8,000 cells (Williams et al., 1998; Strom and Williams, 1998).

At birth, there are three times more retinal ganglion cells than in adults (Strom and Williams, 1998). The massive loss of neurons during the first three postnatal weeks thus parallels that in the ovary. Of much interest to our inquiry, there are statistically strong but partial correlations among inbred mouse strains in the numbers of neurons present at birth and in adults, with correlation coefficients (r) of 0.81 for the 10 strains. Since an r of 0.81 indicates that 66% of the variance in adult cell number is explained by the (genetic) differences among strains (the fraction of variance explained is calculated as r^2), it follows that more than a third of the variance in adult cell number is *not* explained by genetic control on neuron numbers. Strom and Williams (1998) comment: "the unexplained variance must result from strain differences in the severity of cell death, developmental noise, and technical error." We would expand this conclusion by noting that the stochastic feature of cell death (section 3.2) tends to produce a distribution of values, even if the initial number of cells were identical. These mouse strains also differ in extent of cell death and neurogenesis (Strom and Williams, 1998).

Next we consider the hippocampus, a more complex brain structure with dense layers of neurons, which is crucial for declarative memory. Figure 1.5A shows a sample from the granule layer of the dentate gyrus and CA1 pyramidal layer of the hippocampus and the overlying subiculum. The CA1 neurons have particular importance, because they are vulnerable to damage during Alzheimer disease and stroke (ischemia), which cause impairments of memory. The human brain specimens (Fig. 1.5C) showed no evidence of strokes, Alzheimer disease, or other neurodegeneration (West, 1993). Counting error estimates for neuron numbers, determined by independent sampling of the same tissue specimen, are about the same (0.05–0.1%) in hippocampal regions of human (Šimić et al., 1997, West, 1993) and rat (West et al., 1991). These error estimates are small relative to the individual differences in neuron numbers and similar to the retinal cell variations (see above).

In the rat hippocampus and subiculum, the neuron number variation shows a 20–40% range in young individuals, with *C-VAR* of 10–19% calculated from West et al. (1991). In humans, the range of individual values is up to 50% of the mean at all adult ages, indicating about twice the variability in the inbred rat. Similarly wide variations were observed in a different sample of eight adult humans (Šimić et al., 1997). These data suggest that neuron numbers in the hippocampus and subiculum of inbred rats are roughly threefold more variable than in mouse retinal ganglion cells. We wish to keep the species carefully distinct in this discussion, because there may be qualitative differences in the developmental noise among species as well as among brain regions. The greater variability of neuron number in the human hippocampus of presumably unrelated individuals than in inbred rats may represent genetic variations, as described above for retinal ganglion cells. Tentatively, the existing data

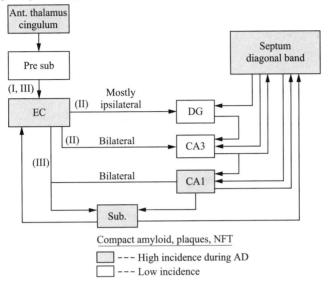

A Hippocampal anatomy

Compact amyloid, plaques, NFT

--- High incidence during AD

--- Low incidence

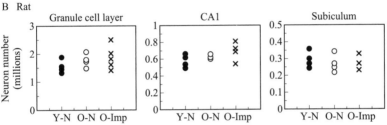

B Rat

Y-N: young (3 mo), normal; O-N: old (24 mo), normal; O-Imp: old (24 mo), Impaired spatial memory

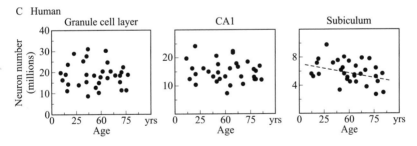

C Human

Figure 1.5. The hippocampus and neuron number variations. *A*, Schema of principal connections within the hippocampus and its relays to other brain regions through the entorhinal cortex (EC), anterior thalamus and cingulum, presubiculum, subiculum, and septum–diagonal band. The numerals in parentheses represent different layers of neurons in the EC. Within the hippocampus, the dentate gyrus (DG) (or granule) neuron layer receives a massive projection from the EC; the DG projects to the CA3 pyramidal neuron layer, which projects to the septum and to the CA1 pyramidal neuron layer.

(continued)

may be interpreted as evidence for up to 40% range of individual differences in neuron number in the hippocampus.

Brain imaging data are consistent with this range in numbers of hippocampal neurons. Of particular importance are data on MZ twins obtained by magnetic resonance imaging (MRI) of living and awake subjects. MZ twins at middle age showed up to 20% differences between co-twins in the total hippocampal volume, as presented as a scatter plot for 10 twin pairs (right plus left side; Fig. 1.6A) and as differences between co-twins (Fig. 1.6B). The MRI analysis has a test–retest (intra-rater) reliability estimate of 0.92 (Bigler et al., 1997), which is as strong as that for retinal ganglion counts (see above). From these extensive individual variations, Plassman et al. (1997) concluded, "MZ twin hippocampal volumes are no more similar than those of unrelated individuals." Moreover, Figure 1.5A shows that co-twin variations in hippocampal size approximate the variations of neurons in unrelated individuals. The variations in left versus right hippocampal volumes are also an example of fluctuating asymmetry (section 1.5.3). There are no direct neuron number estimates of twin hippocampus or other brain regions, but the data strongly suggest appreciable variations in neuron numbers between co-twins and asymmetries with each twin that compare to the variations among inbred rodents.

The corpus callosum, a thick bundle of myelinated neurons that forms the major connections between the cerebral hemispheres, also differed by up to 26% between MZ co-twins (Oppenheim et al., 1989). Although the size of the corpus callosum may approximate the numbers of nerve fibers,

Figure 1.5 (continued)
The CA1 projects to the subiculum (sub.). The shaded boxes represent regions where there is preferential neurodegeneration during Alzheimer disease, in association with a high incidence of neuritic plaques with compact amyloid and neurons containing neurofibrillary tangles. Redrawn from Finch and McNeill (1993). *B*, Rat hippocampal neuron numbers of inbred Wistar rats sampled across the mean life span did not change in three regions. The 2-year-old rats are shown as two subgroups according to whether learning performance in a Morris water maze was normal (O-N) or impaired (O-Imp) as a test for spatial memory. At all adult ages, rats had similar and overlapping numbers of neurons in the hippocampal dentate gyrus and CA1 pyramidal neuron layers, and subiculum, which exceeds the counting error of < 1% (West et al., 1991). Redrawn from Rasmussen et al. (1996). *C*, Human hippocampal neuron number across the mean life span did not change in two regions, dentate gyrus and CA1 pyramidal layer, but did show a loss in the subiculum equivalent to 20% average loss across the life span. These brains were from apparently unrelated individuals without evidence of neurological impairments. The extensive range of neuron numbers at all ages exceeds the counting error of < 1% (West, 1993). Redrawn from West (1993).

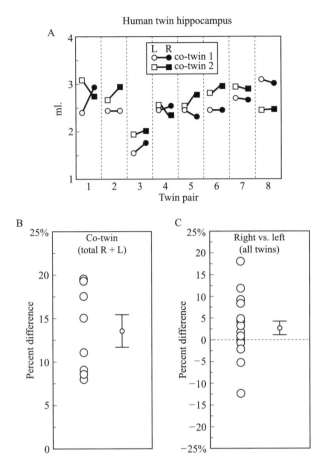

Figure 1.6. The volume of the hippocampus differs within identical (MZ) co-twins. From Plassman et al. (1997); Brenda Plassman of Duke University kindly provided the individual values. Data are not normalized for brain size. Twin pairs 1, 3, and 4 carry apoE ϵ4 alleles, which does not have an obvious effect. The imaging data had a reliability of 0.92 (intrarater reliability coefficient), as established by Erin Bigler in Bigler et al. (1997), a co-author with Plassman et al. (1997). The plane of imaging for these measurements was coronal, in which white matter tracts form the boundary of the hippocampus; there are some uncertainties in the anterior extension of the hippocampus, which may appear to be fused with the amygdala (B. Plassman and E. Bigler, pers. comm.). The calculations were done by Chris Anderson (USC) and C.E.F. *A*, Scatter plot showing that hippocampal volumes of each side vary independently, range of ±20% about the mean. The difference between the left and right sides has the same range of individual variations across all twin pairs. *B*, The percentage difference between co-twins for total hippocampus volume: scatter plot and mean ±SEM of the absolute value of
$[(R + L)_{\text{twin}_1}]/[(R + L)_{\text{twin}_1} + (R + L)_{\text{twin}_2}] \times 100$. *C*, The percentage difference of right versus left side, $(R - L)/(R + L) \times 100$, for each individual across all twins: scatter plot and mean ±SEM.

the number of myelin layers around each neuron could vary among individuals (there are no cell number data).

The cerebral cortex (neocortex) of humans, which connects to the hippocampus through the entorhinal cortex and subiculum (Fig. 1.5), also shows a twofold range of individual variations in neuron density. Pakkenberg and Gundersen (1997) analyzed brains from 94 Danes who died from causes other than neurodegenerative disease or from conditions that were not expected to alter brain functions. Cortical neuron numbers ranged widely among individuals, with a *C-VAR* of 19%, which was threefold more than for body height, that is, ninefold greater variance. The estimated counting error was 3%. Half of the total variance in cortical neuron number was associated with the smaller body size of women; otherwise, body height and weight did not account for individual differences in cortical neuron number. Lacking data on MZ twins, we cannot ascertain the role of genetic polymorphisms in these individual variations.

Next we describe the extensive variations in the patterns of the cortical surface foldings, the *gyri*, which are separated by deep creases, the *sulci*. The neurons within particular gyri have typical roles in major brain circuits, for example, primary visual (striate) cortex, which receives neural input relayed from the eyes. However, the positions of the sulci do not consistently demarcate regional boundaries that correspond to physiological functions (e.g., Whitaker and Selnes, 1975) or to cytoarchitectonics (subregional differences in neuronal morphology; Hustler et al., 1998). Moreover, the adult brain retains much plasticity in the cortical representation of peripheral input, as shown by the large-scale reorganization of cortical maps after accidental and experimental peripheral nerve injury (Kaas, 1991; Davis et al., 1998).

Twin brains have extensive variations in gyral organization and in the size of bilateral structures as resolved by imaging techniques. In part, this information comes from efforts to identify causes of MZ twin discordance in schizophrenia, Alzheimer disease, and other neurological disorders. Twin studies clearly show genetic influences on total cerebral cortical volume, and the sizes of individual gyri and corpus callosum, which are more alike in MZ than in DZ pairs (Jouandet et al., 1989; Tramo et al., 1995; Loftus et al., 1993; Biondi et al., 1998). Nonetheless, Bartley et al. (1997) concluded from MRI analysis of cortical gyri in MZ and DZ twins that cross-correlations were low ($r < 0.25$) and that "most of the variance appeared to be determined by random environmental factors."

A striking example is an MZ twin pair in which the cingulate sulcus is split into four components in both sides of one twin (Fig. 1.7, top) but not the other (bottom; Steinmetz et al., 1994). The different locations of gyri and sulci, even in twin pairs who were concordant for handedness, pose major problems for brain imaging. To those not in the field of brain imaging (and those who may need neurosurgery), it is startling to read:

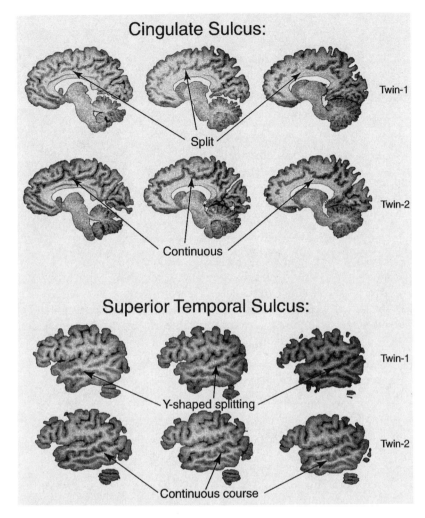

Figure 1.7. A twin pair with different cortical surface morphology: magnetic resonance images (MRI) of cerebral hemispheres, showing that one twin's cingulate sulcus is divided into four parts (*top*), whereas the other twin has the more typical morphology (*bottom*). Both twins were normal in learning ability. From Steinmetz et al. (1994). The print was kindly provided by Professor Helmuth Steinmetz (Goethe University, Frankurt am Mein).

[T]here exists no consensus general cytoarchitectonic map of the human neocortex . . . . The maps set forth by no two cartographers are alike in their detail. Nonconsensus is similarly characteristic of neocortical maps in nonhuman primates. (Caviness et al., 1996)

Efforts to co-register individual variability . . . have revealed tremendous inter-individual variability in the size, shape, and configuration of cortical gyri and sulci . . . [showing] supernumerary gyri, missing gyri . . . and diverse patterns of sulcal branching. (Hustler et al., 1998)

The individualization extends to the total area of cortical regions. Among the cortical regions, the primary visual cortex is particularly variable among individuals, with a threefold range of surface areas as measured postmortem (Gilissen and Zilles, 1996) and correspondingly large variations in the calcerine sulcus, which forms one of its MRI boundaries (Gilissen et al., 1995).

In general, individual variations are greatest in large populations of neurons arrayed in parallel, which in the neocortex exceed $\pm 50\%$ (Williams and Herrup, 1988). The cortex develops through the dispersal of clonally related cells by migration within a generalized protocortical map that does not precisely specify where each neuron will reside or what its connections will be (e.g., Edelman, 1987; Kandel et al., 1992). All of this is consistent with differences in cortical anatomy between MZ co-twins and in the general population that represent chance developmental variations.

We briefly mentioned sex differences above in cortical neuron numbers of humans that showed partial correlations with body and brain size. Moreover, some brain regions have larger sex differences, particularly the "sexually dimorphic nuclei" of the hypothalamus. Figure 1.8 shows major differences among individuals within each sex. In the suprachiasmatic nucleus, which regulates many daily rhythms, males have twofold more neurons than females, aged 10–30 years (Fig. 1.8A; Swaab et al., 1994). Note the 20% overlap in neuron numbers between males and females. Men differ individually fourfold in total neurons and sevenfold in neurons containing the peptide vasoactive intestinal peptide (VIP). Oxytocin-containing neurons of the paraventricular nucleus also vary widely among individuals (Wierda et al., 1991). Although these studies did not employ current stereological techniques or report the measurement error (see above), counting errors are not likely to be large, because of corrections for section thickness by these expert microscopists. Individual variations of neuron numbers in the suprachiasmatic nucleus are about twofold greater than in human hippocampal regions (Fig. 1.5A). In brains from men who died from AIDS, this sample of homosexuals had twofold more neurons than heterosexuals in the suprachiasmatic nucleus, but there is extensive overlap between subject groups (Fig. 1.8).

In sum, these data indicate that neuron numbers vary widely among individuals, but not, in general, as widely as oocyte numbers. Studies of cell counts in the ovaries and brains of the same individuals of several rodent genotypes might reveal if processes of cell generation and cell death are controlled by the same genes. Moreover, data from human twins may inform about whether the greater variation of neuron numbers in humans than in inbred rats represents species differences or genetic variations among individual humans. We note that few journals encourage publication of full sets of numeric data like that from which Figure 1.8 was constructed; however, such data may become more available in website files. In sections 2.6.1 and 4.1, we consider variations of sex

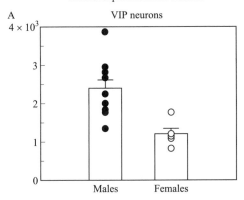

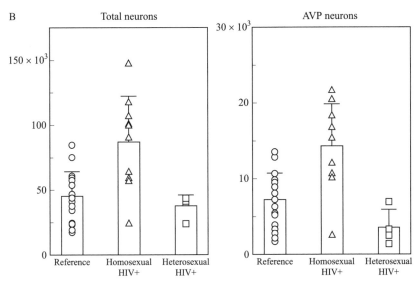

Figure 1.8. Neuron numbers vary more than twofold among individual humans in the suprachiasmatic nucleus (SCN) (specimens from the Netherlands Institute for Brain Research). Calculated *C-VAR* for these samples (C.E.F.): VIP neurons—men (all), 23%; women, 28% (Swaab et al., 1994); arginine vasopressin (AVP) neurons—men (all), 50%; women, 42%. Redrawn from Swaab and Hofman (1990). *A*, The number of vasoactive intestinal peptide (VIP)-immunopositive neurons in males is twofold more than in females, ages 10–30 years. The adult number of neurons is reached by 5 years. The volume of the suprachiasmatic nucleus also varies 3.5-fold among individuals and in correspondence with neuron numbers. From table I of Swaab et al. (1994). *B*, The suprachiasmatic nuclei of homosexual men who died from AIDS had twofold more neurons (total neurons, *left*) or AVP-immunopositive neurons (*right*) than homosexual men who died from other causes or heterosexual men who died from AIDS (Swaab and Hofman, 1990). The AIDS victims were not demented at time of death.

steroid levels in utero that may influence neuron numbers in sexually dimorphic brain regions.

1.4.3. Variations of neuron structures in genetically identical individuals

The complexity of highly specialized cell–cell interactions is commonly understood to be the basis for the enormous range of behaviors which are observed even in simple multicellular animals. The complex question of how synapse formation is determined within neuron circuits can be approached by examining differences among isogeneic individuals.

Synaptic specification in vertebrate brains is widely recognized to have a great range of variations that emerge during development. In an exemplary analysis, Hellwig et al. (1994) examined the spacing of synapses on axons of cortical neurons in mice (NMRI strain, see Table 1.4 note a). The intervals between adjacent synapses fitted an exponential distribution, suggesting that the underlying mechanism governing the spacing of synapses is a stochastic Poisson process in which the lengths of successive intervals are statistically independent. These findings bear on the partly understood mechanisms that determine neuronal connectivity, in which there may be determinism at a global level for synapse density within a circuit, but this determinism does not extend to the precise location of synapses on a particular neuron. We will not proceed further into the large literature on neural circuit specification.

Further evidence for the lack of genetic determinism at a subcellular level is found in genetically identical invertebrates. In *Caenorhabditis*, the positions of some cell types vary widely among individuals, for example, the retrovesicular ganglion (White et al., 1976). Moreover, the branching patterns of each neuron vary, as well as the type of synapses; for example, a particular motor neuron has chemical synpases in one worm but gap junctions in another (Albertson and Thomson, 1976). These variations are still consistent with the generally low "noise level" in the development of *Caenorhabditis* (Ward et al., 1975). Nonetheless, in other nematode species that are built more-or-less like *Caenorhabditis*, there are more common variations in cell number among individuals in certain cell lineages, for example, *Panagrellus* (Sommer and Sternberg, 1995; Sternberg and Horvitz, 1981).

Other invertebrates also show major differences in synaptic architecture among genetically identical individuals (Macagno, 1980; Goodman, 1976, 1978; Williams and Herrup, 1988). For example, in grasshoppers (*Schistocerca*), the location of certain neuron cell bodies in the visual system (ocellar interneurons) can vary more than 200 µm among genetically identical individuals of a parthenogenetic clone (Goodman, 1978). Intriguingly, some clones showed less neuron position variability than others, which indicates that the extent of developmental noise can be influenced by alleles of a few genes that may differ among the clones.

Although gene activity is required at each step of brain development to specify the types of membrane receptors, ion channels, and neurotransmitters in any cell, little is known about how directly *allelic* differences modulate cell position and cell-to-cell interactions.

We do not know from these examples if microscopic variations in cell position or specialized subcellular structures cause functional differences. Although the presence of a chemical synapse versus a gap junction in a motor neuron (see above) might influence the movements or reactivity of an individual, there is every reason to anticipate a larger level of circuit compensation or plasticity. Nonetheless, as suggested in section 1.1, optical methods may resolve how the density of synapses that employ monoamines influences outcomes of aging because this neurotransmitter readily generates free radicals that can damage local cells.

1.4.4. Neuron numbers and the threshold for dysfunctions

Now we consider a model that describes how individual variations in cell numbers may be related to outcomes of aging, drawing on evidence from experimental lesions and Parkinson disease. In section 2.2, we further support this model with experimental studies on how ovarian oocyte reserves influence reproductive aging.

The neuron cell loss required for functional impact appears to differ widely among brain regions, as indicated by a few examples that reveal quantitative relationships between the level of neuron loss and the extent of dysfunction. A very low threshold for neuron loss is observed in certain brain stem neurons. Even the smallest experimental lesions of noradrenergic neurons that regulate blood pressure caused increased pressure fluctuations; these fluctuations increased in proportion to the neuron loss (Talman et al., 1980; Fig. 1.9A). In contrast, another group of neurons in the dopaminergic substantia nigra shows a much higher threshold for neuron loss before clinical symptoms. In Parkinson disease, the onset of movement disorders is associated with a loss of at least 80% of the approximately 400,000 nigral dopaminergic neurons (Bernheimer et al., 1973; Fig. 1.9B). Rat lesion experiments show similar high thresholds for substantia nigra neuron damage to induce movement disorders. The oocyte loss required for menopause is also modeled as a low threshold (section 2.2).

As a working hypothesis, we suggest that cell number variations arising during development in a particular brain region influence the subsequent risk for neuron damage or loss at any age through mechanical injury, tumors, or from age-related neurodegenerative diseases. According to this threshold model, individuals who began life with a higher number of nigral dopamine neurons might be relatively resistant to genetic or environmental causes of Parkinson disease, or show a later onset. The size of the neuron progenitor pool in adult brains (Gage et al.,

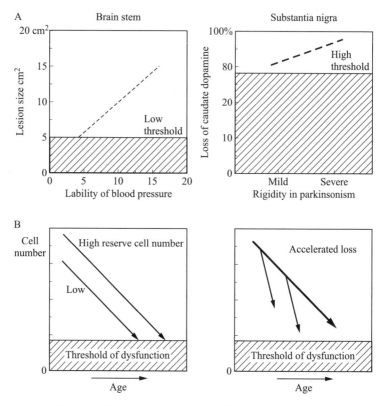

Figure 1.9. Relationships between the size of brain lesions and the threshold for dysfunction in two types of catecholamine neurons. Adapted from Finch (1982; 1990, p. 277). *A: left*, The A2 brain stem nucleus, which contains noradrenergic neurons, has a low threshold (<20% loss) for functional damage. The size of experimental lesions in rats correlates with the blood pressure fluctuations (S.D.). Redrawn from Talman et al. (1980). *A: right*, The substantia nigra dopaminergic projections (A8, A9) have a much higher threshold (>80% loss) than the A2 nucleus, as evaluated in dysfunction in Parkinsonian patients. From data of Bernheimer et al. (1973). *B: left*, The two parallel arrows represent a model for the trajectory of hypothetical loss of neurons during preclinical stages of Parkinsonism. Individuals with higher initial numbers of neurons would reach the threshold for motor symptoms at later ages. The preclinical trajectory of nigral neuron loss is not clear for idiopathic Parkinson disease, that is, that not caused by MPTP or other exogenous toxins. *B: right*, A related model was used to present the hypothesis that the age of onset of Huntington disease results from the interactions of the genetic factor with brain aging processes, in this case accelerating the loss (the downward arrows are at steeper angles; developed from Finch, 1980).

1998) could also be a legacy of development that pertains to this model. (See Fig. 1.9 legend for further background on this model.)

It should not surprise readers by this point that the range of individual variations in the nigral dopaminergic neuron numbers is not known. In general, we do not know much about the individual variability in the numbers of other neuron types and how these variations influence the outcomes of aging and the impact of diseases of aging. Limited evidence suggests that neuron variability and susceptibility to age-related loss are independent traits, for example, as indicated by the different susceptibilities of human hippocampal neurons to age-related loss, despite similar individual variations (Fig. 1.5C).

Of course, specific brain functions depend not only on neuron numbers, but also on the density and biochemical qualities of their synaptic connections, as well as on the plasticity of circuits, including compensatory responses to injury. We may anticipate better understanding of these complex questions through ongoing longitudinal imaging and functional studies of individuals being treated for neurological diseases. The treatments are now targeted at virtually all levels of brain structure and function: neurotransmitter replacement, induction of new synapses, replacement of neurons through the introduction of new cells, and implanted electronic prostheses. Concurrently, many insights will come from mammalian and avian models. For example, studies of songbirds show seasonal changes in neuron numbers in certain brain regions that correlate with the song repertoire (Ward et al., 1998; Tramontin et al., 1998; Nottebohm, 1996; Arnold, 1990). Developmental differences in neuron numbers in some brain regions are hypothesized to modulate adult human sexual orientation (Fig. 1.8B; LeVay, 1991, 1993; Gorski, 1996; Goy, 1988). Until we understand more about the global level of neural circuit plasticity, the variations of synaptic density and of cell numbers are hard to interpret.

We briefly summarize the literature on neuron loss during aging, which remains controversial despite increasingly precise methods. Despite the received wisdom of massive-scale ongoing neuron death, for example, Burne's (1958) famous calculation of 100,000 neurons lost per human brain per day aging, the neuron loss appears to be much more sporadic and disease related than the spontaneous, ongoing loss of oocytes during adult life. It is important to distinguish neuron loss during "healthy" aging from that caused by specific diseases (vascular impairment and stroke; Alzheimer and Parkinson disease). Until recently, there was little assurance that brains from "normal" elderly had been sufficiently screened both premortem and postmortem to identify those with specific neurodegenerative or cerebrovascular conditions of aging. It is clear, however, that no population of neurons vanishes entirely during the human life span in the absence of specific neurodegenerative diseases, which contrasts with the complete loss of ovarian oocytes by midlife in rodents and humans (see Figs. 2.3, 2.4). However, divergent data on neuron loss in the hippocampus are reported even in laboratory rats where the environment and genotype are relatively well controlled (Wickelgren, 1996; Landfield et al., 1996; Gallagher et al., 1996).

In the hippocampal memory circuit, the age-related neuron loss is generally small in the absence of neurological disease (Fig. 1.5B,C). In several studies, the subiculum showed statistically significant neuron loss in humans (West, 1993; Šimić et al., 1997; Fig. 1.5C). The extent of neuron loss in the subiculum and in the CA1 zone of the hippocampus varied in these studies. In long-lived rats (Wistar males), age-related neuron loss in the hippocampus was negligible, even in a subgroup of older individual rats that showed impaired learning (Rasmussen et al., 1996; Fig. 1.5B). Although the age-related subicular neuron loss is small (< 50%) relative to that in Alzheimer disease, it may still be important to age-related memory changes, because the subicular neurons are part of the hippocampal outputs to the entorhinal cortex and the neocortex (Fig. 1.5A). In contrast, significant hypothalamic neuron loss is indicated in specimens of normal humans (see Fig. 2.12). Referring to another neuron type discussed above, retinal ganglion cell numbers also did not diminish at all during aging in rats up to 20 months (Robert Williams, pers. comm.), which is before the mean life span of these strains. We conclude that the extent of neuron loss is modest in the absence of disease and varies among individuals.

1.5. Variations within an individual

Variations in neuron numbers across repeating structures are widely observed *within* individuals. As already encountered in the variations of brain morphology among identical co-twins, these variations are superimposed on the differences in neuron numbers observed between genetically identical individuals. An even more striking example is found in a repeating sensory structure of cold-blooded vertebrates, the lateral line.

1.5.1. The lateral line system

In elegant work reported during 1983–1988 that was an inspiration for this book, Winklbauer and Hausen showed intrinsic randomness in the numbers of lateral line cells at each node within individual larval frogs. These linear arrays of sensory neurons (Fig. 1.10A) detect external motion by a hair (kinocilium), followed by signal transmission to the brain. The importance of the lateral line system to survival (see Fig. 1.10 legend) would lead one to expect that the development of each of its components would be tightly controlled. However, within an individual the numbers of cells in each node of the lateral line vary across more than an eightfold range at this stage in larval development (Fig. 1.10B). Thus, corresponding nodes on each side of an individual can have markedly different numbers of neurons. We discuss in section 3.2.2 how the distribution of neuron numbers within the nodes fits a binomial distri-

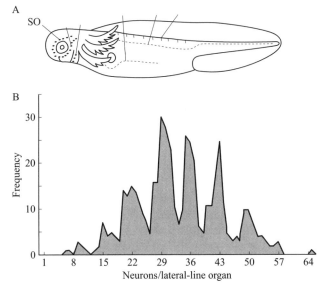

Figure 1.10. The lateral line sensory system. *A*, The lateral line system comprises linear arrays of cell clusters (nodes) around the head and along the sides of the body of fish and amphibia. The supraorbital lateral line in the larval frog (SO) is shown in B; other lateral line systems are indicated by lines but are not named. These cell clusters (also referred to as "stiches" in current literature) contain hair cells (kinocilia) that detect motion on body surfaces, which allows the nervous system to adjust body movements in relation to external movements of water, for example, casued by neighboring fish in a school or by a predator (Cernuda-Cernuda et al., 1996; Smith, 1998; Claas and Münz, 1996). The lateral line also detects water movements of insects and other prey. As terrestrial animals evolved, their descendants lost the lateral line system. However, ciliated cells continue to have key roles in the inner ear to detect sound vibrations in the air and changes in body movement. The brain region of amphibians receiving lateral line information is homologous to the basal ganglia of mammals and birds (Marín et al., 1997). Redrawn from Smith (1996). *B*, Frequency distribution of numbers of sensory neurons in the supraorbital lateral line system of the larval frog (South African clawed frog, *Xenopus laevis*). The frequency distribution of organs for each cell number size class is well described by a binomial distribution (section 3.2.2). Redrawn from Winklbauer and Hausen (1983a).

bution, suggesting a process of random binary choices during development. This example illustrates how chance in cell fate determinism can lead to major differences in cell numbers, even within an individual.

1.5.2. Asymmetry of brain regions

We now return to the bilateral structures of the mammalian brain, which are commonly larger on one side than the other. However, these asymmetries do not always follow the asymmetry in handedness; for example, language functions are typically located in the left hemisphere, which is contralateral to the right hand, but may be ipsilateral (Geschwind and Galaburda, 1984a,b,c). Much data from normal adult co-twins show that the cortical gyri can differ in shape and size between brain sides (Steinmetz et al., 1994; Tramo et al., 1995; Loftus et al., 1993), in some cases by up to 40% between sides (Jouandet et al., 1987). Similarly, the asymmetry of the planum temporale, which is associated with speech and language functions, fluctuated between MZ co-twins (Fig. 1.11A; Steinmetz et al., 1995; Steinmetz, 1996). In each of these twin pairs who were concordant for handedness, the locations of gyri and sulci differed between co-twins. Charles Darwin himself anticipated that "deviations from the law of symmetry would not have been inherited" (Darwin, 1868, vol. II, p. 12).

In the hippocampus of right-handed co-twins, no side was consistently larger than the other (Plassman et al., 1997; Fig. 1.6A). The differences between the left and right sides are in the same ±20% range of individual variations across all of the eight twin pairs (Fig. 1.6C), and exceed estimates of measurement error (see Fig. 1.6 legend). The different hippocampal sizes between identical twins (discussed above) and their asymmetries are plausibly due to cell numbers.

Co-twin differences are detected during fetal development. Ultrasonic tomography studies show that the width of the cerebral ventricles differed by an average of 15% for MZ twins in the second trimester, with further increases in the third (calculated from Table 1 of Gilmore et al., 1996). The planum temporale manifests asymmetry in the second trimester (reviewed in Steinmetz, 1996). The sulci that develop first in the brain (e.g., sylvan sulcus) appear to be more similar in co-twins than the later developing subdivisions, for example, superior temporal gyrus (Tramo et al., 1995; Loftus et al., 1993). The pattern of variance during brain development is not well characterized for most brain systems. In the retinal ganglion, the $C\text{-}VAR$ for cell numbers at birth was not much different from that of adult mice (Strom and Williams, 1998).

In addition to chance developmental variations that arise intrinsically within a fetus, there may also be interactions with fetal neighbors (section 4.1.2). Considerable differences in brain and other organs are found among the genetically identical co-quadruplets of the armadillo (*Dasypus novemcinctus*; Prodohl et al., 1996). Among co-quadruplets obtained just before birth, brain weights varied twofold, suggesting different cell numbers, while brain norepinephrine concentration varied up to sixfold (Storrs and Williams, 1968). Longitudinal studies are needed to show if these major developmental differences persist into adult life. Human co-twins with smaller body size and head circumference tend to catch up in most cases (Keet et al., 1986), but the extent of catch-up growth in neuron numbers is unknown.

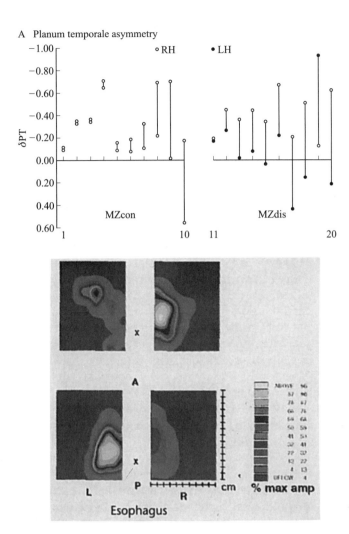

Figure 1.11. Identical twins (MZ) differ in brain asymmetry. *A*, The planum temporale, a language-related area on the posterior temporal gyrus, shows discordant asymmetry in 6 of 10 MZ twin pairs who were concordant for handedness. Among MZ twin pairs discordant for handedness, the left-handed twins showed no lateralization (Steinmetz et al., 1995). *B*, Topographic brain maps for the control of swallowing muscles in MZ twins (Hamdy et al., 1996). Each co-twin was right-handed. However, each co-twin used a different side of the cerebral cortex to control the esophagus. In contrast, they both used the same hemisphere for control of the pharynx. These images were obtained by transcranial magnetic stimulation (TCMS) and are repeatable for an individual. Co-registration of the TCMS and MRI maps was used to define the corticofugal pathways for the esophageal and pharyngeal efferents in the premotor cerebral cortex. In twins and other subjects, hand preference does not predict which cerebral hemisphere controls the swallowing muscles. From Hamdy et al. (1996), with permission.

Having shown individual variations in the volume of different brain regions of twins, it is of interest to consider possible differences in the functional circuits of the brain. A unique insight comes from a study, which fortuitously included an MZ twin pair, of brain regions that control swallowing (Hamdy et al., 1996; Fig. 1.10B). These co-twins were both right-handed but used different sides of the cerebral cortex (hemispheric asymmetry or lateralization) to control the muscles of their esophagus. In contrast, both used the same hemisphere to control the pharynx. In twins and in the other subjects, hand preference does not predict the hemispheric asymmetry for each of the swallowing muscles.

The apparent lack of genetic control on fine structure details of brain circuits is consistent with stochastic models of circuit development though competitive inhibition. Edelman (1987, p. 58) hypothesized that variations in neuron number are a source of individual variations in neural circuitry, including cortical maps. Thus, even if co-twins had identical numbers of cells in their brains at birth, most neuroscientists would expect that each twin would develop different circuits for processing the same specific information. Functional imaging of identical twins could provide rigorous tests of this working hypothesis.

There is evidence for strong genetic influences at higher levels of brain function, in the cortical electric activities detected on the scalp by the electroencephalogram (EEG) and the evoked potentials (Boomsma et al., 1997). In brief, the resting EEG in studies of MZ twins and in family pedigrees shows a very high heritability, explaining 76–89% of the variance. The heritability represented additive genetic factors and differed by EEG frequency (van Beijsterveldt et al., 1996; Anokhin et al., 1992). The amplitude of the P300 evoked potential (300 msec latency) had a smaller heritability of 50%, which was partially associated with different alleles of the dopamine D2 receptor (Noble et al., 1994). Thus, even if the fine structural details of neural circuits (cell numbers, synaptic density) are not strictly specified by the genome, there may be more global features of brain circuit plasticity that compensate in yet unknown ways and that show higher heritability. We do not know mechanisms by which the nervous system maintains coherence at the level of larger circuit interactions, despite the evidently huge level of microanatomical fluctuations.

Differences in neuron numbers of co-twins might account for many examples of discordance in disease onset and incidence. For example, in Swedish MZ twins, Alzheimer disease was discordant in onset by an average of 9.5 years over a range of 4–16 years (Fig. 1.12A; Gatz et al., 1997), in agreement with several other studies. The usual model (Eq. 1.1) would attribute the co-twin differences in onset to environmental differences. In view of 20% differences in co-twin hippocampal size (Fig. 1.5B) and the threshold model of Figure 1.9B, we suggest that a co-twin with later onset was developmentally favored with more hippocampal neurons, which provided a greater reserve and enabled normal mem-

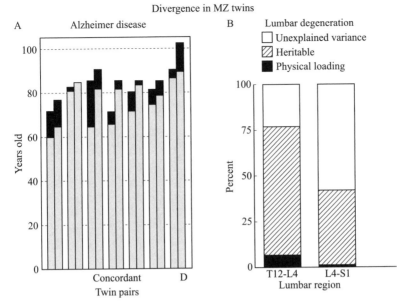

Figure 1.12. Discordances in degenerative disease of co-twins. *A*, Identical Swedish twin pairs who eventually developed Alzheimer disease before death varied by an average of 4.5 years in age of onset between co-twins, over a range of 1–16 years. Redrawn from Gatz et al. (1997). *B*, Lumbar disk degeneration in Finnish MZ and DZ twins (MRI measurements of disk bulging and narrowing) shows spine region differences in unexplained variance. A small part of the variance (< 10%) between co-twins was associated with physical work loads experienced through occupation and recreation. Unexplained variance in the extent of degeneration is twofold greater for lower lumbar disks (L4–S1) than for the upper disks (T12–T14; 55% vs. 24%, respectively). Redrawn from Battié et al. (1995).

ory to persist at later ages during Alzheimer neurodegeneration. Longitudinal brain imaging might test this hypothesis.

Beyond these examples, many other studies of MZ twins show discordances that could arise from chance outcomes in body architecture during development. Nonprogrammed developmental variations in neuron or synapse number could also be a factor in deviant behavior, including discordance of schizophrenia in MZ co-twins, which, like many other psychiatric disorders, has a heritability of < 50% (McGue and Bouchard, 1998; section 5.2). These examples extend to other organs. For example, spinal disk degeneration, while showing heritability in MZ and DZ twin comparisons, could not be predicted by occupational or recreational work load, leaving more than 50% of the variance unexplained for lower lumbar disk degeneration (Battié et al., 1995; Fig. 1.12B). Some divergent patterns of aging in nonneural tissues could be traced back to the nervous system, for example, the greater bone loss in a co-twin who used tobacco than in a co-twin who did not smoke (Hopper and Seeman, 1994), might in turn reflect divergent co-twin tendencies for addiction. Another book could be written to summarize the many fascinating examples of co-twin divergence in aging patterns that are consistent with a limited heritability in outcomes in aging.

1.5.3. Asymmetry and its fluctuations

Asymmetry arises during development at several levels. At the cellular level, there is asymmetric division, in which a cell divides into two daughters with different developmental potentials (Jan and Jan, 1998). The mechanism of asymmetric division may involve extrinsic or intrinsic factors, or both. Extrinsic factors may include contacts with other cells, or contact with molecules in the extracellular matrix, including diffusible signals. Intrinsic factors may involve preexisting asymmetry in the subcellular architecture, or possibly random inhomogeneities in the intracellular distribution of molecules. Well-known examples of asymmetric division arise in the mother–daughter division of the budding yeast *Schizosaccharomyces cerevisiae*, where the daughter forms as a smaller bud from the mother cell whose age, in terms of cell divisions, may be tallied from the accumulated number of bud scars (Sinclair et al., 1998; Jazwinski, 1996). Asymmetry of cellular division in *S. cerevisiae* has particular significance in the context of cell aging, since in this organism the number of divisions that the mother cell may undergo is finite, whereas the newly formed daughter cell has its "age clock" reset to zero, by still unknown mechanisms. A different type of asymmetric division is found in the stem cells of the blood-cell-forming (hemopoietic) system, where a mother stem cell generally gives rise to one stem cell daughter and to one differentiated daughter cell, the latter undergoing further divisions and differentiation events to become a specialized cell such as a T-lymphocyte. The key questions are what molecular mechanisms determine asymmetric cell division and how much randomness there is in cell fate assignment.

At a higher level, asymmetry in the position of certain organs is a regular feature in the body plan of animals, despite the fact that most show a general external symmetry about the midline of the body.

Many internal organs, such as the heart, stomach, spleen, and liver of vertebrates, are intrinsically asymmetric with respect to the left and right sides of the body. Even the simple nematode *Caenorhabditis* has domains of internal asymmetry in which the linear gonad and intestine cross over each other in the middle of the worm. In this case, asymmetry can be traced back to initial asymmetric cleavage of the egg, as is characteristic of nematodes. Switching the positions of two cells through micromanipulation of six-cell nematode embryos reversed the symmetry of adult intestine and gonads (Wood, 1991). This manipulation shows that asymmetric cleavage of the egg initiates a sequence of cell–cell interactions that depends on which cell surfaces of the blastomeres are exposed to each other. Spontaneous reversal of asymmetry is otherwise rare in *Caenorhabditis*.

A different manipulation reversed the location of the heart in the frog *Xenopus*, which, like that of all vertebrates, is on the right side. This experiment perturbed the extracellular matrix of the early embryo by transplantation, or by treatment with heparinase or with RDG tripeptide (arginine-aspartate-glycine) (Yost, 1992). The molecules responsible for the initial asymmetry remain unknown, although many downstream components of the signaling cascade are known, which result in asymmetric transcription of the genes *nodal*, *Sonic hedgehog (Shh)*, and other members of the TGFβ superfamily (Lohr et al., 1997; Ryan et al., 1998). In vertebrates, the highly conserved gene *Pitx2* may be a key downstream transcription target that mediates left–right asymmetry (Ryan et al., 1998).

The brain imaging studies of identical twins discussed above showed that asymmetries in the size of bilateral structures can vary independently of handedness in identical co-twins. Variable asymmetry is widely found in bilateral traits throughout the body that do not show a stable side preference in that species. Given that paired structures develop and grow with significant autonomy from each other, the fact that major asymmetries are not more frequent than they are is evidence of considerable (but not unlimited) genetic control. The burgeoning literature on "fluctuating asymmetry" (FA), as this type of variation is called, mainly examines external anatomic traits (Palmer and Strobeck, 1986; Parsons, 1990; Møller and Swaddle, 1997; Lynch and Walsh, 1998, pp. 113–116). Table 1.5 and Figure 1.13 illustrate FA in the lengths of appendages and in numbers of insect bristles. An example of FA from the lab of C.E.F. is the numbers of ova shed by the right versus left ovary in inbred mice, which has a dextral bias, as does the uterine decidual response (Holinka and Finch, 1978); for both parameters, the dextral bias fluctuated over a range from equivalence to > 50% among individual mice.

Morphological traits differ widely in the amount of FA. For example, the FA of eye-facet number and wing length in the fruitfly and for mouse whisker number is small relative to FA of fly microchaetes (Mather, 1953; Dun and Fraser, 1959; Soulé and Cuzin-Roudy, 1982). As discussed for brain asymmetry, the FA of some traits is independent of the internal asymmetry of visceral organ alignment (left–right handedness, coiling of the intestines, location of the heart, etc.). In general, the amount of FA within an individual is much less than the variations among individuals (Soulé, 1982; Soulé and Cuzin-Roudy, 1982; Leamy, 1986).

Of particular interest, there are no correlations between different body traits showing FA, which implies that FA arises as localized developmental fluctuations. For example, after artificial selection of mice for increased or decreased widths of the first molar (M_1), the resulting FA of M_1 was not correlated with the FA of M_2 (Leamy, 1986). Fluctuations

Table 1.5. Anatomical traits showing fluctuating asymmetry (FA)

Species	Organ (number of units)	Measure of FA[d]	References
fruitfly	Bristles		
	sternopleural chaetae, 9[a]	FA_1, 1%; FA_3, 16%	Mather, 1953
		FA_1, 3–6%	Jokela and Portin, 1991
	scutellar microchaetae, 18[b]	FA_1, 1%	Usui and Kimura, 1993
	Eye facets (ca. 100)	FA_1, 5%	Waddington, 1960
mouse	Molar width (M_1, M_2)[c]	FA_1, 1%	Leamy, 1986
	Humerus length	FA_2, 1–15%; FA_3, 4–21%	Siegel and Doyle, 1975
	2° whiskers (vibrissae; 19/cheek)	FA_1, <1% (wild-type)	Dun and Fraser, 1959
human	Ear breadth	FA_1, 3–5%; FA_3, 7–10%	Kobyliansky and Livshits, 1989

[a] Numbers of sternopleural chaeta (macrochaetae, large bristles) vary less than the microchaetae. The degree of FA may be correlated with the numbers of bristles in the breeding line (Jokela and Portin, 1991).

[b] Numbers show microchaetae in row 1 of the scutellum + prescutellum; $N = 18 \pm 1.7\,SD$ (Usui and Kimura, 1993). Data were not available to calculate the FA.

[c] FA as a proportion of total variance, 16% (Leamy, 1986).

[d] Because the methods of data analysis vary among studies, it was not possible to give the same parameter for each example. Mouse and fruitfly were inbred. Corrections for directional asymmetry are included in mammalian traits. For bristles and other meristic traits that represent repeated anatomical units, we also show the typical number of units. Calculations of FA are discussed by Palmer and Strobeck (1986) and Møller and Swaddle (1998). Notation adapted from Livshits and Kobyliansky (1991):

FA_1, grand mean of the absolute value of (left–right)/mean size.

FA_2, $(1 - r^2)$; r is the correlation coefficient for left and right size (Siegel and Doyle, 1975; Van Valen, 1962).

FA_3, $C\text{-}VAR$ of differences between the sides of FA_1.

may arise during development within the context of local tissue domains. For example, tooth germs in mammals, and imaginal disks of insects have cellular boundaries that may contain fluctuations locally and limit interactions with those in other domains.

Fluctuating asymmetry in adults may be traced back to variations in early development, which establish the architecture of asymmetry, and which, through successive events in cell differentiation, lead to local variations of symmetry in internal and external organs and entities. It is therefore of particular interest that rare individuals, about one in 10,000 in humans, have the condition *situs invertus*, which results in a complete reversal of the right–left handedness of these asymmetric organs. *Situs invertus*, one imagines, would be an unwelcome surprise to a surgeon who had failed properly to examine a patient before operating, for example, the appendix would be on the wrong side. Individuals with this condition are generally without other clinical symptoms, and the

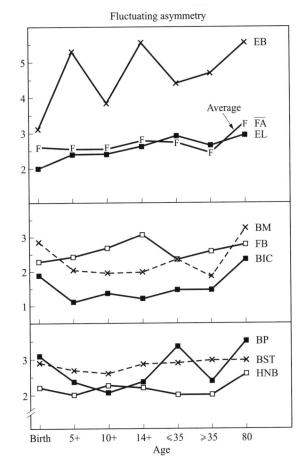

Figure 1.13. Fluctuating asymmetry of eight bilateral traits increased with age in normal humans: $\overline{FA}$, mean fluctuating asymmetry, calculated as FA_1 (Table 1.4, notes); EL, ear length; EB, ear breadth; BIC, bicondylar breadth; BM, bimalleolar breadth; FB, foot breadth; BP, biepicondylar breadth; BST, bistyloid breadth; HNB, hand breadth. Cross-sectional data of males (M) and females (F) are from an Israeli population. Age group (number of subjects): 18–29 years (N = 424 M, 1038 F); 36–45 years (93 M, 59 F); 80–85 years (28 M, 76 F). Redrawn from Kobyliansky and Livshits (1989), with permission.

inverted organs appear to function normally. In mice, two mutants, *iv* and *inv*, were identified that specify the right–left handedness of organs (Yokoyama et al., 1993; Afzelius, 1996; Lowe et al., 1996). The *iv* gene does not itself appear to specify a particular handedness, but mutants homozygous for the defective *iv* allele are equally likely to be right- or left-handed, whereas *inv* mutants are always inverted. The mechanism in *iv* involves a mutation in dynein, a microtubule-based motor, which is expressed in early development (Supp et al., 1997).

However, even when major organs are on the proper side, details of internal anatomy can still vary in normal individuals. Lack of determinism is conspicuous, for example, in the branching airways of the lungs and of the microvascular networks that carry blood to tissues. The topology of networks of microvessels is typically asymmetric and irregular, and the lengths and diameters of the vessel segments making up the networks are highly variable (Pries et al., 1996; Fig. 1.14). The branching number for human blood vessels—that is, the number of bifurcations from a larger to a smaller vessel—has been estimated to be 32 for the arterial system and 25 for the venous system (Morimoto, 1998). The resulting fine-grained heterogeneity, which can be expressed in terms of fractal properties (Gazit et al., 1995; Baish and Jain, 1998), arises from the need to adapt to growth and varying local functional demands. In the human brain, vascular anatomy varies individually even more than patterns of gyri and cytoarchitecture (section 1.4.2) (surveyed by Whitaker and Selnes, 1975). These authors point out that loss of blood flow through a given arterial branch, such as through occlusion or aneurism, could have widely different impact on function, depending on the individual variations in vascular anatomy and in the localizations of brain circuits. Angiogenesis at lower levels is presumably governed by local factors to allow adaptive responses to demand. Studies on twins and inbred mice might show the level at which random variations in the pattern of blood vessels become clearly detectable.

Flucuating asymmetry has much interest among evolutionary biologists, because of indications of its correlation with individual fitness (Møller and Swaddle, 1997). For example, the FA of wing length was greater in flies eaten by birds, which implies that higher FA makes a fly easier prey (Møller, 1996a). Other examples in mammals are more suggestive than conclusive. Extreme asymmetries in facial features are indications of mental or physical disorder and therefore negative in mate attraction (Shackelford and Larsen, 1997). However, some degree of facial asymmetry may be a component of attractiveness (Swaddle and Cuthill, 1995). A subgroup of human subjects with extreme right frontal lobe activation had a selective, if modest, deficit in the natural killer (NK) class of lymphocytes (Kang et al., 1991), whereas paw preference in rats was associated with higher NK (Neveau, 1993; also see the Geschwind-Bahan-Galaburda hypothesis, section 2.6.1). Furlow et al. (1997) analyzed the performance of college students on Cattell's culture fair intelligence test (CFIT). A combined measure of FA based on nine body (traits (finger length, ankle breadth, ear breadth, etc.) was inversely correlated with CFIT scores. However, only a minor portion of the variance was associated with FA (4%, $r = -0.2$; $p < 0.025$), and other behaviors did not correlate with FA (Møller and Swaddle, 1997). Because FA may be increased by stress during development (section 5.3.1), further studies of different human populations may be revealing, especially those with

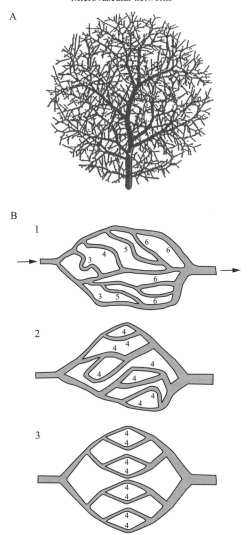

Figure 1.14. Models for analysis of heterogeneity in microvascular networks. *A*, Computer-generated model of a vertebrate vascular system using fractal geometry. From Williams (1997), with permission. *Bottom* (from Pries et al., 1996), with permission: *B.1*, Simplified diagram of an observed vascular network with irregular morphology and topology, with each microvessel labeled by its generation number in terms of branchings from left to right. *B.2*, Model of a symmetric irregular network in which each microvessel has the same generation number, but segment diameter and length vary. *B.3*, Model of symmetric regular network in which each microvessel has the same generation number, segment diameter and length.

higher exposure to environmental risks and with a higher incidence of learning disorders.

A further question is the extent to which FA in adults results from chance effects during the adult life span in parts that undergo regular remodeling. For example, skin and connective tissue throughout the body undergo continual turnover, with replacement of epidermal cells within several weeks and slower replacement of extracellular matrix molecules. Chance injury or exposure could locally perturb cell turnover and cause deviations that might switch the direction of asymmetry. This outcome of aging appears consistent with data on FA on humans aged 18–85 years, in which the *C-VAR* of many traits and the degree of FA of ear dimensions and other characters increased progressively during aging (Livshits and Kobyliansky, 1989, 1991; Fig. 1.13). The degree of FA did not correlate between individual traits, suggesting that the differences are independent of one another, as would be expected for locally autogenous processes. The age-related increases of FA imply that higher values of FA do not predict early mortality, contrary to the evolutionary hypothesis that the degree of FA indicates fitness (Møller and Swaddle, 1998).

Because cross-sectional data can be biased by death of impaired individuals, longitudinal analyses are needed to evaluate the persistence of FA in individuals. Two examples of longitudinal studies give different outcomes that illustrate the range of mechanisms at work. Barn swallows showed persistence of the side with the longest tail feather during 12 years of observed seasonal molts (Møller, 1996b). The story is more complex for claw length asymmetry in crabs. In the brachyuran crab *Hemigrapsus nudus*, claw asymmetry is retained after molting (Chippendale and Palmer, 1993), whereas the hermit crab *Clibanarius vittatus* follows the chirality imposed by the host shell (Harvey, 1998). In *Clibanarius*, the asymmetry does not depend on initial possession of a shell, suggesting that the morphology has a genetic component. However, if individuals are reared without shells, the asymmetry can be lost within as few as five moults, indicating that environnmental cues are important for the maintenance of the asymmetry. Longitudinal studies of FA may reveal different patterns of stability according to local molecular and cellular turnover and exposure to external fluctuations.

1.6. Variations in molecular controls

1.6.1. Pattern formation and the suppression of errors

A remarkable feature of development is the reliability with which complex structures develop under genetic control. By "reliability" we mean that supernumerary or missing structures are rare, and that the sizes of organs and other structures show well-defined ranges of variations for each species. Because the force of natural selection is much stronger on early life stages than on outcomes of aging (section 1.7.3 below), it is not surprising that developmental variations are far more constrained than outcomes of aging. There is much evidence for sophisticated mechanisms, which were evolved to provide fail-safe pathways to reliable development that limit the extent of errors in patterning during development beyond a particular tolerance or boundary value (Cooke, 1998; Namba et al., 1997; Wolpert et al., 1998). Here we examine a particular instance of pattern formation—the emergence of somites in the developing chick embryo— which illustrates some of the mechanisms through which pattern emerges and which illuminates the challenge of understanding how intrinsic chance variations are constrained, if not entirely suppressed.

In vertebrate development, segmentation of the mesodermal cell layer of the early embryo gives rise to the somite series, which then imposes a segmental character on other organ systems. Details of somite formation vary from species to species, but the general organization is similar. The initial somite segments are developmentally equivalent units that form in a head-to-tail sequence through the successive formation, at regular times, of regularly spaced fissues that subdivide previously uniform columns of tissue. The control of somitogenesis is global in the sense that, even though individual embryos of a species may differ in extent by as much as 35%, the species-characteristic number of somites has a variability of 5% or less (Cooke, 1998). Typical numbers of vertebrate somites are of the order of 40–50 units between the rear of the skull and that of the pelvic articulation. Individuals with a deviance of 1 unit may occur as frequently as 10%, whereas a deviance of 2 units may approach a frequency of 1%; interestingly, there is as much variation in relative counts within the subregions of vertebrae of different types (neck, thorax, lumbar, etc.) as in the overall count (Jonathan Cooke, pers. comm.).

Elegant work with early chick embryos by Palmeirim et al. (1997) has revealed some aspects of the mechanisms by which somites may be formed. Somite formation proceeds along the track of the regressing primitive streak, a strip of cells that is the forerunner of the head-to-tail axis (Fig. 1.15). Bilateral blocks of tissue, called segmental plates, form initially containing the cells that will form the first 10–12 somites. The first pair of somites then develops at the head end, and the segmental plates extend at the tail end by equivalent amounts. After 90 minutes, the second pair of somites has formed and the segmental plates have grown further. This process repeats itself at regular intervals so that a wave of somite formation passes down the body axis. The segmental plates maintain a constant length of 10–12 somites worth of cells during most of this process—in effect, the segmental plates are a sliding window of developmental activity.

The chick segmental plates display an extraordinary phenomenon of rhythmic cycles of expression of the gene *c-hairy1*. This putative transcription factor is a homolog of the *hairy* gene, which is a mediator of

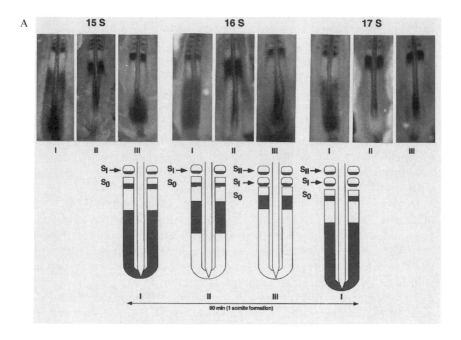

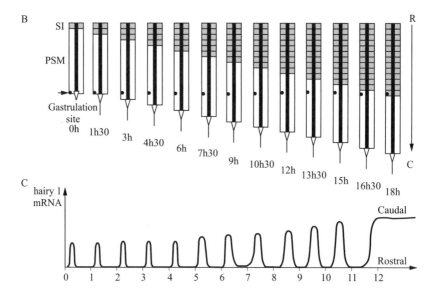

Figure 1.15. *A*, In situ hybridization showing patterns of *c-hairy1* mRNA expression during somite formation in avian embryos aged 15, 16, and 17 days. Schematic representation of the pattern of *c-hairy1* mRNA expression in the presomitic segmental plates with the progression of somite formation.

(*continued*)

segmentation in fruitfly embryos (see below; Palmeirim et al., 1997). Within each cell of the segmental plates, there are 12 cycles of synthesis of *c-hairy1* mRNA lasting 90 minutes. Between successive cycles the mRNA becomes undetectible. The cycles propagate as wavefronts of activation that start at the tail end of the segmental plates, each new wave beginning as the previous one reaches the head end of the plates. Furthermore, the waves narrow and probably intensify as they pass from the tail to the head end of the segmental plates. As each new somite forms, it freezes its *c-hairy1* expression pattern in such a way that its head end is "off" and its tail end is "on" (Fig. 1.15).

The degree of coordination and control suggested by these data illustrates the complexities of developmental patterning. In a simple model, each cell within the segmental plates is subjected to a series of sequential and graded signals resulting in the "waves" of *c-hairy1* gene expression, each successive wave being shorter and sharper than the one before it. The final wave throws a switch in the tailmost strip of cells within the next-to-form somite, which blocks *c-hairy1* activation in the head end of the new somite and results in a precise segment boundary being formed. The molecular control of these processes is not known, but appears to require both an "oscillator" to set in train the successive waves and a precise intracellular mechanism to detect and respond to a graded series of cycles (Cooke, 1998). We do not know whether the signals are intrinsic to the segmental plate cells, or whether they are cued by some intercellular paracrine factors. Either case is likely to involve extensive interactions between adjacent cells, such as observed during lateral signaling in bristle pattern formation in *Drosophila* (Heitzler and Simpson, 1991; Kopan and Turner, 1996). In short-term organotypic culture, the segmental plate cells, once started, support autonomous cycles of *c-hairy1*

Figure 1.15 (continued)
Stages I, II, and III correspond to the formation cycle of one somite. In stage I (left) S_I denotes the most recently formed somite. While a new somite (S_0) is forming from the upper end of the segmental plate, a narrow strip of *c-hairy1* expression is seen at the lower end of the nascent somite, and a large expression domain extends from the tail bud region (stage I; *A, D, G*). As somite formation proceeds, as evidenced by the appearance of a fissure, the large band moves upward and narrows (stage II; *B, E, H*). When the new somite is almost formed, the band has narrowed considerably (stage III; *C, F, I*). Finally, when the new somite is completely formed, the cycle begins again. *B*, The segmental plates (white bars) contain about 12 prospective somites and migrate downward, adding new cells from the gastrulation site (Hensen's node and rostral primitive streak) and leaving the column of somites above them. *C*, Between the moment a new cell joins the segmental plate and the time it is incorporated into a somite (18 hours), it will see 12 waves of *c-hairy1* expression. From Palmeirim et al. (1997), with permission.

expression (Palmeirim et al., 1997). However, some external signal, perhaps from the regressing primitive streak, is required to initiate the correct spatial organization of cycling.

Although the molecular mechanisms of pattern specification may prove to vary widely among organ systems and phyla, this example of periodic waves of at least one mRNA during somitogenesis clearly implies both the intrinsic potential for errors and the presence of sophisticated mechanisms for modulating variations in cellular signaling that enable the reliable emergence of patterning. Mathematical models indicate how chance fluctuations in numbers of signaling molecules would, if uncorrected, give rise to disruptive levels of errors in some development pattern-forming events (Lacalli and Harrison, 1991; Holloway and Harrison, 1999). Therefore, of great interest are the mechanisms by which embryos suppress the potential errors that could otherwise arise.

1.6.2. Control of gene expression

In the cell nucleus, the production of messenger RNA by each gene is determined mainly by the rate of transcription by the RNA polymerases. The specificity of effects on genes arises from the relatively tighter binding of transcription factors to response element sequences than to neighboring DNA, typically by 10^3–10^4. In many subcellular mechanisms, small molecules such as cAMP or larger molecules like transcription factors must diffuse over distances of many microns through the cytoplasm into the cell nucleus. Two stochastic features occur in this process. First, there is the action of chance in determining the actual timing of a ligand–receptor interaction, once the ligand and receptor become colocalized within a sufficiently small space to allow stereochemical interactions ("ligand-gated" receptor activation). Second, one (or both) of the interacting molecules needs to diffuse, sometimes over considerable distance in relation to its size, so that the first of these processes can come into play ("diffusion-limited" signaling). The latter case is characterized generally by longer delays, as the diffusion time may be on the order of seconds or minutes whereas interaction times may be measured in milliseconds or less. These movements involve vast numbers of molecular collisions by Brownian motion (random walks) in three dimensions before reaching the site of the DNA in the promoter of the gene that may be responsive (see section 1.1). The initial binding to DNA may be distant from the highest affinity location that, directly or indirectly, regulates RNA polymerase activity. The protein may then diffuse by sliding along the DNA helix until a higher affinity site is reached. In many examples, gene regulation requires the assembly of multiple transcription complexes each of which includes diverse factors. The concentrations of each of these molecules must be regulated in relation to the dissociation rate from the DNA target, to assemble the required suite of transcription regulators within the time requirements of the cell system. The statistics of these diffusion-limited processes are not well described and may be very complex. Obviously, gene regulation cannot and does not depend only on a single fortuitous collision of a single transcription factor with one high-affinity site of one gene (see below). Other rate-limiting signaling is mediated by external ligands that bind to receptor proteins on the cell membrane. Membrane-binding ligands can be small molecules, such as steroids or small peptides, or can be large multichain proteins. Again, the efficacy of membrane receptor activation depends on binding the appropriate ligand more tightly by 1000-fold or more than other proteins. Through random walk collisions, a large number of unsuccessful contacts may occur before the ligand is bound by high-affinity sites in its target receptor.

Just as in the assembly of transcription factor complexes, each step in the synthesis of RNA and protein molecules also depends on diffusion-limited arrival of the individual nucleotides and amino acids from which these macromolecules are constructed. Gaspard et al. (1998) confirmed that Brownian motion of colloidal particles (2.5 μm diameter) suspended in water is essentially chaotic, the microscopic dynamics of the system being very sensitive to its initial state. Similar behaviors may be expected for diffusion of smaller molecules. Each of these interactions proceeds at rates that represent specific probabilities based on the rates of atomic and molecular movement.

This intrinsic noise in molecular motion could be a factor in many aspects of cell fate variation and could play a role in fluctuating asymmetry (Soulé, 1982). A glimpse of these processes at a molecular level is given during development of mechanosensory bristles in fruitflies which are organized into rows (Hartenstein and Posakony, 1989; Schweisguth et al., 1996; Usui and Kumura, 1993; Simpson et al., 1999). Within each nascent row of bristles, there is no fixed schedule for emergence of fine bristles (microchaetes) at a particular location in a row, as observed by a cell-type marker (*A101*) during a phase of 4–12 hours (Usui and Kimura, 1993). Moreover, the number of microchaetes that eventually forms varies between individuals (Usui and Kimura, 1993; *C-VAR* in female fruit flies is 5%, calculated from Hartenstein and Posankony, 1989). These variations contribute to the fluctuating asymmetry of fruitfly bristles (Table 1.5). Underlying these processes is complex cell–cell signaling which involves the Notch pathway (Fig. 3.1) and multiple events in gene circuits that eventually lock-in cell-type specific patterns of gene expression. It is hypothesized that signaling fluctuations lead to the selection of one cell from a group of neighbors to become the bristle precursor cell (Kopan and Turner, 1996; Simpson et al., 1999).

Mathematical modeling of a simple prokaryotic gene circuit (McAdams and Arkin, 1997) (Fig. 1.16, 1.17) illustrates the intrinsic noisiness of biosynthetic processes during the flow of gene information, which are implied in the above example of variations in bristle pattern formation. The simple circuit with one gene and one protein product that inhibits its own synthesis (negative autoregulation by a dimer) is more typical of bacteria and their viruses than multicellular organisms, but the

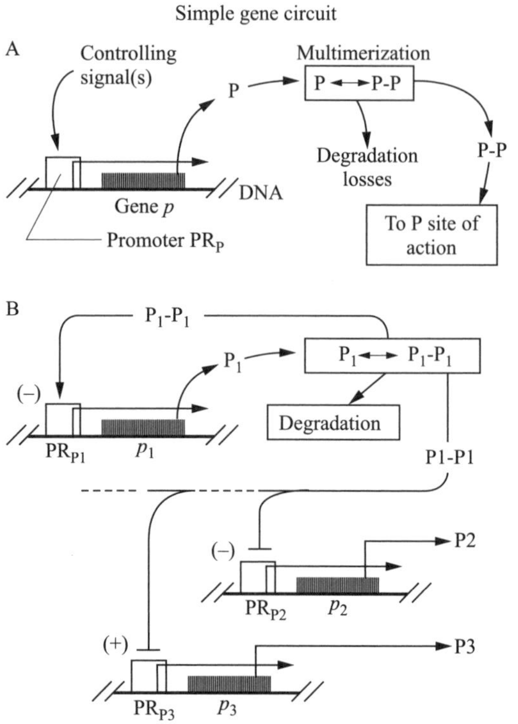

Simple gene circuit

Figure 1.16. System for control of gene expression based on the stochastic model of McAdams and Arkin (1997). *A*, A promoter *PRp* controls transcription of gene *p*, which results in protein product P. Protein P undergoes dimerization prior to acquiring functional activity. *B*, The model of *A* is extended to allow for autoregulation by P_1 dimers of gene p_1 expression. P_1 dimers also control expression of genes p_2 (PRp_2) and p_3 (PRp_3). From McAdams and Arkin (1997), with permission.

principles that emerge have general relevance. In this model, the gene product forms a dimer that feeds back to regulate its own promoter. Promoter regulation by dimers and more complex protein assemblies is typical of eukaryotic gene systems, in which transcription and translation are not simultaneous and occur in different subcellular compartments with specialized diffusion and transport characteristics, and feedback systems at multiple levels that constrain the extent of variations. Moreover, nucleated cells have 10- to 100-fold more active genes and multiple compartments with different diffusion and transport characteristics. Eukaryotes also have highly evolved feedback and scaffolding systems to constrain the extent of random perturbations in biosynthesis and signaling that would otherwise cause maladaptive variations, as discussed further below.

McAdams and Arkin (1997) used the approach of Gillespie (1977, 1992) for a stochastic simulation of coupled chemical reactions that

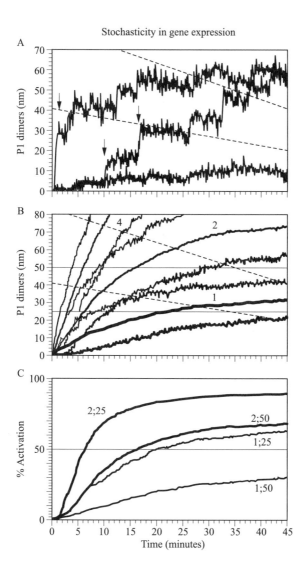

Figure 1.17. Simulations from the stochastic model of McAdams and Arkin (1997). *A*, Three simulation runs using identical parameter values for production of P_1 dimers in model B of Figure 1.16. The pattern of protein expression shows differences between simulation runs. Abrupt jumps (arrows) represent chance high fluctuations in output from bunching of short transcript or translation intervals. *B*, At three p_1 gene dosage levels (1, 2, 4), 100 simulations were run. The central curves at each gene dosage level show the mean rise in P_1 dimer level, and the curves above and below show mean ±1 SD. Variation in the time to reach critical thresholds for P_1 dimer level can be seen to decrease with a more rapid rise in P_1 dimer concentration, associated with higher gene dosage. *C*, Mean activation level of gene p_3 promoter for different values of $N; K$, where N is the P_1 gene dosage and K is the dimer:promoter binding constant. For each curve the activation level in individual cells will vary widely about the mean, as in B. From McAdams and Arkin (1997), with permission.

pertain to organisms at all levels of complexity. This model differs from conventional assumptions of chemical kinetics, in which changes in a reaction system over time are considered continuous and deterministic. In the stochastic formulation, a reaction *probability* per unit time corresponds to the macroscopic reaction *rate* parameter. Monte Carlo simulations can be used to evaluate the time course of the coupled reactions. The assumption that reactions are probabilistic, rather than deterministic, can be seen as representing the random-walk nature of the molecular movements by diffusion and the chance nature of ligand–receptor interaction, although these fine-scale events are not themselves explicitly represented in the model. In the model of Figure 1.16 distances over which the molecules need to diffuse are short and thus the model is principally one of ligand-gated receptor activation. Transcript initiation is modeled as a single reaction, with an exponential distribution of time intervals between successive transcripts. As the newly formed RNA polymerase moves downstream of the site of initiation by one-dimensional (vectorial) diffusion, the cleared space allows the next polymerase to load. Intracellular concentrations of gene regulatory components are initially low. As each transcript emerges in this prokaryotic model, it is attached by a ribosome and a peptide chain is initiated, also subject to stochastic influences. From the probability of each discrete event in a reaction sequence, the algorithm calculates the resulting numbers of each molecular species and, by accumulating the results for all reactions over time, gives an estimate of the statistical fluctuations in protein product P_1.

Figure 1.17 shows three computer runs with different outcomes in the levels of P_1 dimers formed over time. Each run developed abrupt upward steps and declining phases due to the probability distributions of the duration of each step of molecular assembly and degradation. The fluctuations in regulatory protein concentration were large relative to the threshold concentrations that feed back on gene promoters. The time required to reach the critical threshold concentration for transcription feedback is equivalent in machine terms to the switching delay.

Models of this type provide a valuable insight into the chance variations seen in cell behaviors. For example, the bacterium *Escherichia coli* reveals considerable variation in the cell generation time within a growing culture (*C-VAR* of 22%) of genetically identical organisms in a homogeneous environment. Although relatively tiny in size (1000th of the volume of a mammalian cell), the bacterial cell cycle nonetheless also depends on the diffusion-regulated activities of hundreds of genes. We may expect extensive noise in processes of gene expression whenever the concentration of the regulating ligands is low, as is the general case for transcription factors and other signals that mediate variable gene expression and cell differentiation. Feedback systems to check runaway drift from intrinsic small, random perturbations in biosynthesis are implied (Emlen et al., 1993). We argue that switching delays due to the statistics of diffusion are key contributors to chance features of cell fate assignment and, in turn, to developmental variations in the numbers of cells and cell units.

Other data on gene regulation show features consistent with the above analysis that helps explain how noise creeps into gene information flow. The rates of these signaling processes depend directly on the concentration of transcription factors that must be assembled on the multiple regulatory modules of each gene. The inventory of different transcription factors varies among cells, but is probably in the range of hundreds of different proteins that bind to and regulate the promoters of the many thousands of different genes that are typically active in differentiated eukaryotic cells.

Eukaryotic gene regulation is characterized by multiple regulatory modules, the gene promoters, that may be contiguous and upstream of the initiation site of transcription, with multiple transcription factors that bind in each module. The inventory of regulatory regions includes domains with positive and negative effects on transcription (enhancers and suppressors, respectively) of one or more genes. For individual genes, the specificity of cell type expression depends on particular subsets of this inventory, which in some well-studied cases is about five different transcription factors per regulatory module (there may be several regulatory modules in each promoter, each of which can have multiple independent binding sites for the same transcription factor) (Arnone and Davidson, 1997). The required concentrations per nucleus for DNA binding are determined primarily by the binding affinity constant to the DNA site and are typically hundreds to thousands of each type of DNA binding transcription factor (Harrington et al., 1992; Cutting et al., 1990). The assembly of transcription factor complexes may also depend on the combinatorial association of multiple proteins. Arrival at their target DNA sequence by diffusion may be guided by scaffolding proteins in the nuclear matrix (Nickerson et al., 1995) (section 1.1). Cooperative interactions between homodimers or heterodimers may increase DNA binding affinity and interactions with yet other proteins. DNA response elements can compete for limiting concentrations of transcription factors. Moreover, in some genes the modules are separated by hundreds or thousands of nucleotides, which requires an additional diffusion step to align two distant assemblages of DNA–protein complexes.

In each of these steps, the arrival times of transcription factors at the DNA binding site are determined by diffusion-limited processes that would be expected to yield probabilistic distributions like those incorporated in the McAdams and Arkin (1997) model of a much simpler (prokaryotic) system. The kinetics of forming these structures are not known in detail but are probably on the order of minutes. The processing of the initial RNA transcript within the nucleus and its diffusion out to the endoplasmic reticulum also takes some minutes. Presumably, the duration of these diffusion events scales with molecular size. Although the time distribution of these processes is not yet described, one may model it as an exponential distribution, right skewed with a long tail, as modeled for transcript initiation intervals in prokaryotes (McAdams and Arkin, 1997). Indications of randomness in transcriptional activation were described in macrophage cells stably transfected with an HIV-LTR promoter and a *lacZ* reporter, which showed expression in a subpopulation of cells, which increased with the strength of external stimuli (Ross et al., 1994). These cell–cell variations in gene expression recall the indications of variable expression of the *hsp-16* gene between individual genetically identical nematodes (Section 1.2). Few studies have been designed to evaluate the *range* of quantitative variations in the constitutive expression of a particular gene within differentiated cells of higher organisms. This information cannot be simply obtained from conventional in situ hybridization, because of variations in cell geometry. It is possible that the quantitative distributions of protein markers used to discriminate cell subpopulations in fluorescence activated cell sorting (FACS) (e.g. Bix and Locksley, 1998) also represent quantitative differences in gene expression.

Arkin et al. (1998) extended their analysis of stochastic effects in reaction rates affecting control of gene expression to examine the outcomes of random molecular-level fluctuations on the lysis-lysogeny decision of bacteriophage infections. This well-characterized competitive regulatory mechanism is a powerful model for studying how an initially homogeneous cell population can undergo partition into distinct subpopulations following different cell fate pathways. The random choice between the lysogenic or lytic path in individual cells is consistent with molecular-level thermal fluctuations in the rate-determining reactions of gene expression mechanisms. Arkin et al. (1998) concluded:

> This analysis indicates how molecular level thermal fluctuations can be exploited by the regulatory circuit designs of developmental switches to produce different phenotypic outcomes. Such regulatory mechanisms will produce diverse phenotypes even in clonal cell populations maintained in the most homogeneous laboratory environments. In such systems environmental signals can act on the parameters of the regulatory circuits to bias the probabilities of path choice under different conditions. (p. 1645)

Future approaches to modeling eukaryotic gene circuits which includes stochastic features might reveal how genes implement preprogrammed cellular development, while changing cell processes and structures in response to environmental signals and intrinsic noise. In the near future, we anticipate knowing much more about how multicellular eukaryotes deal with the intrinsic stochastic variations in molecular movements during gene expression, macromolecular biosynthesis, and cell-cell interactions. Constraints on stochastic fluctuations are provided in part by scaffolding proteins at cell membranes and in the nuclear matrix (Ranganathan and Ross, 1997; Nickerson et al., 1995; Mastick et al., 1998). These and other subcellular organizations serve to guide the diffusion of molecules and may well affect the precision of molecular interactions. Much more will be learned about the kinetics of assembly of transcription factors, which are postulated to involve successive molecular displacements in which stochastic fluctuations may determine the outcomes of cell differentiation (Enver and Greaves, 1998; Sieweke and Graf, 1998) (section 3.2.4). Extensive feedback mechanisms at many levels operate throughout development that constrain the variability in alternate paths of cell differentiation and eventually allow the development of patterns and structures with a limited and genetically specified range of individual variations (Emlen et al., 1993).

1.7. Limits to genomic specification

Genetic information determines the architecture of individual organisms during development and programs the mechanisms that maintain function during adult life. The gene pool, as carried in populations of a species, represents the outcome of natural selection for optimum phenotypes for the current (or past) environment. These conclusions, which stand among the great scientific accomplishments of the twentieth century, are still open to an important qualification when aging is considered. Within an individual, genetic information is not transmitted with perfect accuracy to all of the cells of an organism at any age, nor is it expressed with 100% reliability. Mutations in DNA can accumulate, with the result that, with sufficient time, the soma becomes a genetic mosaic. Even if a gene is unaffected by somatic mutation, a fraction of its gene products at any given time will contain errors. During aging, individuals typically accumulate changes that depart from the evolutionary optimum, leading to impaired reproduction and increased mortality at later ages. These tendencies are reinforced by the declining force of natural selection during adult aging. Moreover, evolutionary theory predicts trade-offs that favor fitness at earlier ages at the expense of fitness at later ages (see below).

1.7.1. Programmed versus stochastic aging

By saying that chance affects a process or event, we do not necessarily mean that the outcome is *entirely* random. In fact, it is difficult to establish the degree of randomness (Pincus and Singer, 1998). Chance and its action tend to be subject to constraints in living systems. We must understand both the nature of the constraints and the forces generating randomness within the bounds of these constraints.

In aging research there has long been a polarization between the so-called "programmed" theories versus the "stochastic" theories (Kirkwood and Cremer, 1982). The stochastic theories have been criticized on the basis that aging cannot be truly random because it is too reproducible a phenomenon. But "stochastic" does not mean completely random. Stochastic simply means that chance plays a key role, almost certainly within the constraints set by the genome. That is, there is *pro-*

grammed stochasticity, at least in the sense that the genetic program sets the parameters of the statistical distribution of phenotypic values. This view is compatible with the general canonical pattern of aging in mammals, with vulnerabilities of particular organ systems (Table 1.2) and the differential expression of the gene encoding telomerase, which strongly influences the extent of telomere DNA loss during cell replication (section 1.7.2).

Considerations of the evolutionary basis of aging (see also section 1.7.3) show that it is unlikely that any program specifies the end game of life in detail for the simple reason that the force of natural selection at old ages is too weak. This contrasts sharply with the processes of development and maturation, when the force of selection is at its apex. Development, particularly in the early embryo, as all would agree, is genetically programmed in the fullest meaning. But even so, developmental programs allow some range of chance variations.

Each day, each of our cells experiences a substantial number of random oxidative hits to its DNA from endogenous free radicals. By some estimates, the steady state number of modified bases is around 20,000 per cell in rodents, although the range of uncertainty is large (Beckman and Ames, 1998; Helbock et al., 1998). Most of the damage will be repaired, but some persists as somatic mutations. Once again, which DNA lesions are repaired and which escape repair is thought to be largely random. There is therefore a daily chance that a somatic mutation will occur in a vulnerable cell that will be the final, triggering step in initiating cancer. The overall contribution to the aging process from damage by free radicals and other oxidants like glucose is thought to be extensive and to be a shared element of aging across a wide phylogenetic range (Martin et al., 1996). Because reactive oxygen species are inevitable by-products of oxidative phosphorylation, on which most cells depend for energy production, the cellular antioxidant defenses, which include enzymes such as superoxide dismutase, catalase, and glutathione peroxidase, are well conserved across the species range. However, there is evidence that long-lived species invest in higher levels of antioxidant protection than short-lived species (Kapahi et al., 1999), indicating that genetic systems can and do evolve to constrain the effects of chance at the molecular level. A good example of how the constraints operate differently within different species may be seen in the case of spontaneous incidence of malignant tumors. The highly efficient human DNA repair processes ensure that on the whole cancer is rare until relatively late in life, when in the "wild" we would probably be dead already from natural external hazards. This constraint on the action of chance is determined by our genotype. While mice get cancers at much younger ages than humans, the lifetime risk for a mouse of developing a tumor is about the same as for a human (Ames et al., 1985). The event—initiation of a tumor—is in both species random; the *distribution* of events is constrained.

In anticipation of ensuing discussions on how chance operates during development, we outline the levels of biological organization in which chance events are observed. At the atomic level in biological organization, there are many types of chance interactions of energy with molecules. For example, photons in the ultraviolet range of sunrays may hit skin cells with sufficient force to damage their DNA and cause mutations. Over a lifetime, these cells, or their clonal descendants, may acquire further hits and, if not killed, may progress toward malignancy.

Molecular diffusion is another fundamental process that involves the action of chance (section 1.6.2). Biomolecules of all sizes are constantly in a state of Brownian motion through random walks at living temperatures. Because the rate of Brownian motion decreases with increasing particle size, large molecules like proteins will move less on this account than do small molecules. Steroid hormone signals also typically bind to large proteins, the steroid receptors, which can take many minutes to find their way by diffusion from the cytoplasm into the nucleus of a cell. Growth factors in the sticky external matrix environment of cells may also diffuse quite slowly.

Cell migration during development is directed by signals from molecules in the extracellular matrix. Direct observations of migrating cells, as will be described for the developing lateral line system (section 3.2.2), show fluctuations indicative of guidance mechanisms that constantly assay local molecular cues. During the migration of cells in the developing ovary and brain, some are left behind as ectopic cells, which have a shortened life span (section 3.1.3). Cell death is also subject to chance as an outcome of cell differentiation. During the extensive proliferation of cells in vertebrate and insect embryos, the fates of the daughter cells at each division are determined by chance, so there can be a wide range in numbers of cells of a particular type, depending on whether a cell death program is activated in a particular daughter cell. We will argue that the variations in cell numbers, which we described above for oocytes and neurons, are the result of chance processes in cell fate determination during development.

During cell replication, the duplicated chromosomes separate and line up on the metaphase plate, followed by their separation into the two daughter cells. The complex mechanics that effect these processes are subject to a molecular instability that allows for errors. Occasionally a chromosome strays behind, so an extra chromosome is added to one daughter cell and lost from the other (aneuploidy). We do not know the actual rate of aneuploidy in vivo because dysfunctional cells are selected against. While aneuploidy is often lethal in normal somatic cells, it can occur in the oocyte during the last meiotic reduction just before ovulation. Although most aneuploid embryos die, humans with an extra copy of chromosome 21 frequently survive to develop Down syndrome. This risk increases with maternal age, with an impact on aging of the offspring, because trisomy 21 causes premature Alzheimer disease (section 2.4).

The environment has innumerable influences on development, which also influence outcomes of aging. We may consider a fetus unlucky if it grew in a mother who used addictive drugs or who had a suboptimal diet. In chapter 4, we will consider these and other examples of chance operating on the fetal environment, including interactions between neighboring fetuses, which may be of different sex. Once born, individuals are obviously at risk of accidents from extrinsic risk factors such as predators, infections, famines, and so on. The duration of life is clearly much influenced by these essentially random factors.

Random as they are, most of the accidents and other extrinsic challenges faced by organisms conform to statistical patterns, which forms the basis for natural selection and adaptations in the form of antipredator defense, immune defenses, and so on. For example, the overall level of extrinsic hazard to which a population is exposed is the major factor in determining the rate of attrition by mortality in a cohort of same-aged individuals, and this rate of attrition determines how the force of natural selection becomes attenuated in older animals (Rose, 1991; Charlesworth, 1994). The decline in the force of natural selection during aging is the basis for evolutionary theories of aging, as described below.

Organisms have also evolved strategies to cope with fluctuation in food supply. For example, undernutrition in utero and in the early postnatal period can permanently change the body structure and physiology, causing increased risk of cardiovascular disease, hypertension, and diabetes in later life (section 4.4). Harmful effects of fetal undernutrition are suggested to result from an adaptive response that spares the growing brain at the expense of other organs, resulting in unbalanced growth. In a contrary example, the health of laboratory rodents may be improved and life span lengthened by *caloric restriction* after puberty, that is, controlled underfeeding by 10–40% below the ad libitum level (Sohal and Weindruch, 1996; Masoro and Austad, 1996). Caloric restriction increases life spans (by up to 50%) and delays progression of many canonical features of aging cited in Table 1.2. Dietary-restricted rodents show normal body proportions but reduced fat depots, and they have enhanced repair mechanisms and resistance to physiological stress. Some of these beneficial effects extend to primates (e.g., Edwards et al., 1998), but effects on primate life span are not yet known. The response to caloric restriction is hypothesized to represent an evolved adaptation to cope with intermittent food shortage, with temporary shutdown of reproduction and investment of relatively more resource in maintenance, in order to retain health for reproduction when conditions become more favorable (Holliday, 1989; Harrison and Archer, 1989; Masoro and Austad, 1996; Shanley and Kirkwood, 1998). The plausibility of this hypothesis depends critically on the average frequency and duration of periods of food shortage, and therefore on the pattern of action of chance.

1.7.2. Errors in gene expression

Inaccuracies in the expression of genetic information arise in the DNA strand itself, in the messenger RNA made on DNA templates in the nucleus and mitochondrion, and in the proteins encoded by this RNA.

The error rate of DNA is very low—estimated at one mistake per 10^7 to 10^8 DNA bases—because of powerful enzymatic error correction mechanisms that act to maintain fidelity. Transcriptional error rates are two to three orders of magnitude greater, for example, around one mistake per 10^5 nucleotides (Anderson and Menninger, 1986), chiefly because RNA polymerases lack the efficient exonuclease proofreading activity that characterizes DNA polymerases (Goodman et al., 1993; Petruska and Goodman, 1995; Moran et al., 1997). The error rate in protein synthesis is variously estimated as around one mistake in 10^3 to 10^4 amino acids (Kirkwood et al., 1984; Rosenberger, 1991). These error rates are all low, when one considers the intermolecular discriminations required in the complex enzymology of biosynthesis at DNA and RNA templates. Low error rates in genetic information transfer are not simple outcomes of "lock-and-key" stereochemical interactions between enzymes and substrates, as implied by Watson-Crick base-pairing rules and genetic code tables. Low error rates result from atomic level interactions that promote accuracy against a background of continual random molecular movements.

The discrimination between correct and incorrect substrates is derived ultimately from differences in the interaction of free energies of correct and incorrect molecular partners in enzymatic biosynthesis, but the final discrimination far exceeds the basic selectivity that is predicted from these primary differences in binding energies. General models proposed some decades ago by Hopfield (1974) and Ninio (1975) suggested how the enzymatic machinery could generate enhanced molecular discrimination through use of high-energy intermediate states within the reaction pathways or by incorporating a delay within the reaction process to allow kinetic amplification of the discrimination afforded by differential dissociation rates of noncovalent interactions with cognate/noncognate substrates (see Kirkwood et al., 1986, for general discussion). These general concepts indicate how the elements of molecular discrimination may be made more efficient, although in the important example of DNA replication it is now clear that the principal means to achieve high fidelity is through aggressive exonuclease proofreading (Goodman et al., 1993; Petruska and Goodman, 1995; Moran et al., 1997). Elegant experimental systems are addressing the exact mechanisms that determine DNA replication fidelity, but as observed by Goodman (1997):

The surface has barely been scratched in terms of understanding the inter-actions between polymerases and DNA that determine replication fidelity. The magnitude and location of mutations depend on a complex interplay between polymerases, proofreading exonucleases, processivity factors, and the properties of DNA primer-template sequences. (p. 10495)

The relative slowness of DNA, RNA, and protein synthesis as enzymatic processes is consistent with the kinetic complexity required for various types of error surveillance and error correction (Ninio, 1987).

Kinetic and probabilistic mechanisms are also important to the control of gene expression and other aspects of molecular signaling (Ninio, 1986; McAdams and Arkin, 1997), including antigen presentation to the T-cell receptor (McKeithan, 1995; Kell et al., 1998; Torigoe et al., 1998). The control of transcription at gene promoter sequences is regulated by combinatorial interactions among several DNA binding proteins. The random diffusion of cell signaling molecules and the combinatorial assembly of multimeric complexes at membrane receptors and in gene regulation engender extensive chance variations. The conversion of cel-lular signals into altered cellular reponses, or "signal transduction," is mediated largely by a set of posttranslational modifications of proteins (phosphorylation and dephosphorylation). The benefit of limiting chance fluctuations in multimeric signaling complexes could have been a factor in the evolution of scaffolding proteins (section 1.1).

The reading of diffusible signals is important not only during devel-opment but also to maintain proliferative homeostasis in tissues that are continually renewed from the division of stem cells (section 3.2). Normally, a stem cell divides asymmetrically, producing one daughter stem cell to maintain a constant stem cell pool, and one differentiated cell. However, if the stem cell pool is depleted, for example, by irradiation damage, both daughter cells may become stem cells. The mechanisms regulating stem cell numbers in the bone marrow, skin, and gut remain stable through long periods. Nevertheless, there is evidence that some of these stem cell homeostatic systems begin to deteriorate with aging. Chance is involved in the determination of cell fate under normal con-ditions, and may also play a role in the alteration of cellular homeostasis in old age.

During protein synthesis, the estimated error rate of one mistake per 10^3 to 10^4 amino acids results in substitutions that will be scattered at different locations throughout the peptide chain of different molecules. Depending upon the size of the protein, this implies that throughout the body a sizable fraction of proteins will have one or more missubstituted amino acids. However, the chance of malfunction from these errors is probably much less, because many amino acids can substitute for each other, and because molecules with altered structures are usually degraded rapidly through proteolytic pathways. The error hypothesis for accumu-lation of aberrant proteins in aging tissues motivated studies in the 1970s and 1980s to detect altered proteins in tissue from aging animals and in replicatively sensescent cell cultures (Holliday, 1987; Sharf et al., 1987; Rothstein, 1982; Finch, 1990, pp. 398–400). However, most molecular geontologists interpret the accumulation of altered proteins with aging as a result chiefly of posttranslational damage from glycation or free radicals, rather than errors in biosynthesis. Damaged proteins are sometimes resistant to proteolysis, which may lead to their intracellular accumulation. Moreover, oxidative damage can accumulate in extracellular proteins such as collagen and elastin, which may be formed before maturation and therefore are nearly as long-lived as an individual.

There are many difficulties in measuring the *randomness* of missubstitutions. If each error per molecule were unique, this could not be detected by the usual techniques for variant peptides, such as two- dimensional gel electrophoresis. Techniques for separating variant proteins depend on segregating distinct subpopulations of molecules that comigrate in a gel or that can be detected by immunological cross-reactions. Even though these techniques are increasingly sensitive to small populations of variant molecules, they still require significant numbers of proteins to be altered in the same way. There is no technique to directly amplify trace amounts of proteins that is equivalent to the powerful polymerase chain reaction (PCR) methods for nucleic acids. Thus, we still know little about the sequence heterogeneity of proteins in vivo and its possible biological consequences.

The somatic genome is also subject to chance variations, particularly in proliferating cells (sections 3.2 and 3.4). Cancer is hypothesized to be a consequence of somatic genome instability resulting from a multistep process of error accumulation, in which errors in DNA replication led to point mutations, or chromosomal rearrangements accumulate in a cell clone (e.g., Lengauer et al., 1998). A major effort is being made to characterize the accumulation of damage to somatic DNA during aging. Markers include the oxidized base 8-oxoguanine (Beckman and Ames, 1997) and chemical adducts to DNA that resemble those induced by carcinogens in vivo (Randerath et al., 1993). Both types of DNA markers show tissue specificity during aging.

The ends of chromosomes (telomeres) contain short repeated DNA sequences that do not code for known genes. The enzyme telomerase maintains these sequences during development and in some adult tissues, including germline cells (Kipling, 1997; Lingner and Cech, 1998; Harley and Sherwood, 1997) and certain lymphocyte stem cells (Weng et al., 1998). When the gene for telomerase is not active, telomeres are subject to random loss of telomeric DNA, which is observed in vivo in some lymphocyte subtypes (Weng et al., 1998; De Boer and Noest, 1998) and in the cell culture aging model of diploid fibroblasts during serial culture (Bodnar et al., 1998). If the telomerase is replaced in cultured fibroblasts, then their telomeres are maintained, together with a greatly increased ability to proliferate (Bodnar et al., 1998). The chance activation of telomerase may also be a step in carcinogenesis, although active telomerase alone does not cause abnormal growth.

Genome instability is also seen in mitochondrial DNA (mtDNA), which accumulates point mutations and deletions in certain tissues during aging in mammals, for example, Lin et al. (1998), Melov et al. (1995), Soong et al. (1992). Adult cell types can differ widely in the prevalence of mtDNA deletions at later ages, even within the same organ; for example, the dopamine-rich basal ganglia accumulate mtDNA deletions, the Kearns-Sayre syndrome (KSS) deletion of 4,977 nucleotides in nearly 0.5% of mtDNA, which is 1000-fold more than in the cerebral cortex of the same older humans (Soong et al., 1992). The cell type location of these mtDNA deletions is not known in the brain. Neonatal brains had no signal for the KSS mtDNA deletion. A variety of other mtDNA lesions have now been described, whose specifics do not concern us. The human myocardium and kidney (Liu et al., 1998) accumulate far fewer mtDNA lesions than does skeletal muscle (Zhang et al., 1998; Kopsidas et al., 1998). In particular, skeletal muscles present a mosaic with an apparently random distribution, in which individual muscle fibers deficient in cytochrome oxidase activity also had greater heterogeneity in PCR-generated mtDNA sizes, indicative of mtDNA mutations (Kopsidas et al., 1998).

The variation among cell types may arise from several factors, each of which includes a chance component. In dividing cells, the mitochondrial population must double each time a cell divides. In postmitotic cells, mitochondria replication is required to keep pace with the loss through turnover, with approximate life spans of several weeks. Mathematical modeling of the intracellular accumulation of defective mitochondria shows that a high mitochondrial division rate may delay the accumulation of mutations (Kowald and Kirkwood, 1993). This is consistent with the observation that the greatest age-associated accumulation of mtDNA mutations is seen in postmitotic tissues (brain, muscle), which have slow mitochondrial division rates. In addition, there may be local tissue differences, such as the greater load of oxidative stress in dopamine-rich brain regions (see above). Systemic metabolism further interacts with mtDNA lesions in aging skeletal muscle, which are decreased in rats by caloric restriction (Aspnes et al., 1997) and are associated with hyperglycemia in aging humans (Liang et al., 1997). These metabolic influences suggest that individual differences in diet and the degree of age-related blood glucose elevations are important background factors in chance damage to mtDNA in some tissues. Thus, we have two examples in which the pattern of gene expression acquired during development modifies the DNA error rate: in the tissue differences of mtDNA mutations and in the expression of telomerase.

In contrast to these largely negative chance variations in DNA, those that arise through recombination in lymphocytes are a key step in acquiring a competent immune system during development and in postnatal immune responses. Recombination is mediated enzymatically by the recombinant activating genes (RAGs). The selection of clones of cells that respond to a particular antigen depends on having an enormous inventory of mutated and recombined immunoglobulin genes available. The RAG activity promotes a high level of site-specific genetic recombination within the variable (V) regions of immunoglobulin genes. Chance operates here on outcomes of aging by the gene combinations that are available for body defenses at later ages, as well as chance variations that cause some of these clones to become autoreactive and induce later autoimmune diseases. Autoimmune diseases with an age-related component include multiple sclerosis, rheumatoid arthritis, polymyalgia rheumatica, and giant cell arteritis, which are rare at young ages. Identical twins who are discordant for clinical signs of multiple sclerosis can nonetheless show degrees of abnormalities in white matter (Uitdehaag et al., 1989), which might arise from chance variations in the generation of autoreactive T-cells, or from differences in brain cell reactions to autoreactive T-cells.

Chance outcomes of cell fate determination are another major limit to genetic authority. As discussed in detail in section 4.1, cell differentiation proceeds statistically, such that at each cell division during development, the daughter cells may acquire different cell fates, which can include programmed cell death or differentiation into different cell types. In general, most somatic cell types cannot be traced to exactly defined lineages in early development. Germ cells are an exception, and their lineage is determined early in development in a wide range of animals, but not all. Thus, genetically identical individuals, depending on the species, can have variations in the numbers of cell of each type (see Figs. 1.3, 1.4, and 1.8). These variations are discussed further in chapters 2–5.

1.7.3. Genes, evolution, and aging

Given the formidable ability of biological systems to cope with chance events during development, it seems at first sight paradoxical that during aging, the robust health of the young adult spontaneously becomes compromised even in optimum environments. Again, there are wide chance variations in the schedules of dysfunctions among individuals carrying the same genes (MZ twins, inbred mice and fruitflies, self-fertilizing nematodes).

Evolutionary theory provides powerful arguments about the genes that affect aging and longevity. Here we expand on the concepts briefly presented in the introductory remarks to this chapter. In essence, two factors explain why aging occurs (Kirkwood and Rose, 1991). First, in the natural world the risk of mortality from extrinsic, natural hazards is typically high. These hazards include predators, infections, food shortages, temperature fluctuations, and injury. This high level of extrinsic mortality means that force of selection to sustain fitness later in adult life is negligible or absent, simply because so few individuals chance to survive to late ages (Medawar, 1952; Williams, 1957; Hamilton, 1966; Charlesworth, 1994). Second, longevity depends on the continued action of maintenance and repair functions, which require energy to work. This means that selection may trim investments in somatic maintenance to be just enough to maintain the soma in good condition through the natural life span in the wild, but not for much longer (Kirkwood, 1977, 1981; Kirkwood and Holliday, 1979). This concept is called the "disposable soma theory." For example, in natural habitats, 90% of wild mice die by 10 months, whereas in the laboratory at least 50% would survive for 24 months (Austad, 1997). With few exceptions, the majority of reproduction in a population is contributed by the young adults.

This demographic situation is predicted to result in accumulation of genes in the population that have delayed adverse effects that comprise aging, and in an evolved limitation in somatic maintenance and repair processes, on the basis that there is no advantage to be gained from investing in better somatic repair than is needed to get the organism through its natural expectation of life in the natural environment. An example of a gene with delayed adverse effects would be one that increases the risk of Alzheimer disease. Such a gene would be selected against if the disease began early in life, but would not be strongly selected against, if at all, if dysfunction began at much later ages when the gene would already have been passed on through earlier reproduction. At least seven different genes have variants associated with Alzheimer disease, consistent with the accumulation of sporadic germ-line mutations during human evolution (section 5.2). Regarding the ovary, selection to continuously form new oocytes through de novo oogenesis in mammals was apparently weak, presumably because the age of reproductive senescence when oocytes are exhausted is late enough to be compatible with the prior accomplishment of sufficient reproduction.

Differences in the rate of decline in the force of natural selection at later ages provide the basis for explaining why different species have characteristic life spans. For example, organisms in high-risk ecological niches are predicted to invest less in somatic maintenance and repair than organisms in low-risk niches, which would select for genes that have differing efficacy in repair of ongoing damage from, for example, free radicals. Of course, no known single gene displays species differences in activity or level of expression that could account for the scaling of aging across the 30-fold range of mammalian life spans. The more likely explanation is that multiple genes regulating somatic cell maintenance and resistance to stress evolved to support higher somatic maintenance required for the prolonged postnatal development typical of longer lived species (Kirkwood, 1992; Martin et al., 1996; Finch and Sapolsky, 1999). Recent findings from the lab of T.B.L.K. support this hypothesis: primary skin fibroblasts cultured from mammalian species with a wide range of life spans showed resistance to oxidative stress that correlated positively with longevity (Kapahi et al., 1999). Such data support the idea that chance processes, such as cell damage following stress, are integral to aging processes.

1.8. Synopsis of the argument

This book is an inquiry about how outcomes of aging may be shaped by the role of chance variations in development, which led us to assemble a new body of evidence from diverse scientific literatures. The motivation for this inquiry is evidence that life span has modest heritability in two mammals (human, mouse) and two invertebrates (fruitfly, nematode). Even within highly inbred animals and identical twins, life spans vary widely. While these observations argue strongly against a role of chromosomal gene variants, it remains possible that life span variations could be due to chance effects from somatic cell genetic instability, such as mitochondrial DNA mutations. We then describe further types of anatomical and physiological variations that are not attributable to ordinary genetic diversity. Inbred rodents differ severalfold in the numbers of ovarian oocytes at birth and in their patterns of fertility cycle changes during aging. The numbers of neurons in inbred rodent brain regions also vary to a lesser degree. These variations in cell numbers may also contribute to nonheritable differences in the symmetry of paired organs (fluctuating asymmetry). Many of the examples introduced in this chapter are examined in more detail throughout the book.

Chapter 2 describes chance developmental events affecting cell numbers in the reproductive system and how these modify individual patterns of aging. In chapter 3 we consider the mechanisms of cell number variation more generally. We return in chapter 4 to the operations of chance in the external environment of the embryo, with examples of effects on individual aging. Chapter 5 develops a general synthesis of these findings and examines the diversity of gene risk factors for Alzheimer disease and certain other age-related conditions that show variable penetrance. We end by making suggestions on how to pursue these questions experimentally and for applications to preventive medicine.

Note

Gärtner (1990) also discussed "a third component causing random variability besides the environment and the genotype." Gärtner and his colleagues had observed extensive variations in body and organ weights of mice that were not decreased by extensive inbreeding (Gärtner, 1990; Baunack et al., 1986), but did not broadly examine these phenomena. In part, these variations may arise from the fetal neighbor interactions in rodents, which are a special source of variability in adult phenotypes (section 4.7.2). Phelan and Austad (1994) review the increased variability associated with inbreeding, which is observed in some, but not all, gene systems.

2

Chance and Reproductive Aging

Chapter 1 introduced the wide individual variations in life spans, reproductive aging, and cell numbers in ovary and brain. We now use the reproductive system as a model for the principles of how developmental variations set the stage for outcomes in aging. We will consider the hypothesis that variations in ovarian oocyte numbers at birth determine the duration of active reproduction in humans. We show how the variations in ovarian oocyte numbers arise during the organogenesis of the ovary because of chance aspects of germ cell differentiation and migration. We point out the wide individual variations in human fetal sex steroid levels during midgestation, when sex steroids have fundamental influences on sex hormone–sensitive neurons. We also show how the prostate is subject to chance developmental variations that may be associated with later life benign prostatic hypertrophy (BPH). Later, in chapter 4, we show how fetal neighbor interactions in rodents and DZ twins can introduce yet other variations in brain development.

2.1. The mammalian ovary

The ovary of mammals is a unique example of an organ that ages irreversibly because the numbers of its cells are set during development. No other organ loses its key cells at such a rate during aging. The adult ovary contains follicles at various stages of growth, each of which surrounds a single oocyte (Fig. 2.1). During the reproductive cycles, which begin at puberty, a tiny fraction of the primary follicles will complete the maturation process to allow ovulation. No one knows why so few primary follicles begin to grow during each ovulatory cycle, while the rest stay dormant, for 50 years or more in humans. During the life span, a female

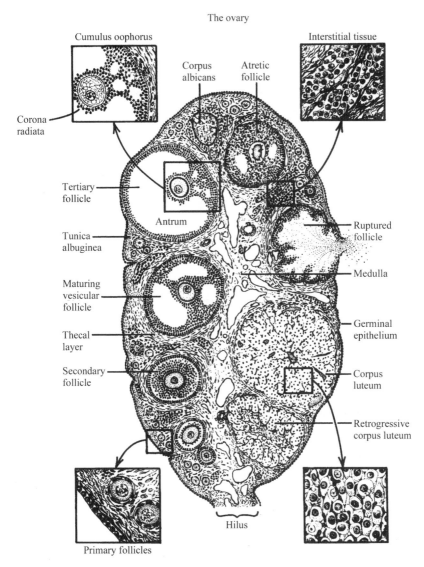

The ovary

Figure 2.1. The ovary of adult women contains follicles of a range of sizes, each of which surrounds a single oocyte. From Freeman (1994), with permission.

mouse may ovulate about 5% of its neonatal stock of oocytes, while a woman may ovulate less than 0.1% of hers. The vast majority die, as described below.

As a follicle grows, its cells differentiate to produce increasing amounts of the female sex steroids estradiol and progesterone. The transitions from primary to mature follicles take about 3–4 weeks in rodents but, surprisingly, are not under strict feedback control from the pituitary gonadotropins (Hirschfield, 1991). Eventually, the estradiol secreted by the larger follicles reaches a critical level in the circulation that triggers the brain and pituitary to release a surge of gonadotropins (luteinizing hormone [LH] and follicle-stimulating hormone [FSH]) into the circulation, which in turn causes the largest follicles to burst and release an ovum (ovulation). In rodents, 6–20 ova are released (the exact number depends on the genotype). Human ovaries usually release a single ovum, except in the infrequent 1% of cycles, when two or more follicles ovulate simultaneously (see Table 4.1). Unless fertilization has occurred, the burst follicle regresses rapidly, with a precipitant drop of progesterone that causes the steroid-dependent uterine cells to slough off, forming the menstrual discharge.

The growing follicle populations can be represented as a pyramid of ascending size classes of growing follicles that include progressively fewer of each larger size follicle (Fig. 2.2A). For each size class the probability can be estimated that a follicle and its oocyte will transit to the next stage of growth or will die (Faddy et al., 1983; Fig. 2.2B). The death of a follicle and the loss of its contained oocyte are together called follicular atresia. How follicles are chosen for further maturation versus atresia is not well understood (Hirschfield, 1991). Death is most frequent in the larger follicles of adults, but also in small follicles of infants which suggests that there is at least one major branch point during follicular development (Faddy et al., 1983; Hirshfield, 1988).

Two possibilities are debated, that continued follicular maturation represents "pure chance" as to which growing follicles die or survive, or that defective follicles are culled out. The preovulatory surge of FSH, which lasts only a few hours in rodents, appears to select a subgroup of follicles that are at just the right stage of growth. It is unclear if other follicles that did not respond are defective. Even at birth, there is evidence for heterogeneity of oocytes. The few oocytes that form in the fetus at the end of gestation in mice are located at the outer edge of the ovary (this small subset was detected by labeling with tritiated thymidine, which was infused at various stages of pregnancy; Hirshfield, 1992). This finding is consistent with the hypothesis of Henderson and Edwards (1968) that oocytes are ovulated in the order in which they were formed during development. Ongoing studies may show whether the late-formed oocytes are ovulated at later ages.

At least some of the follicles and oocytes that are lost can be rescued, as shown by a classic experiment. If one ovary is removed surgically

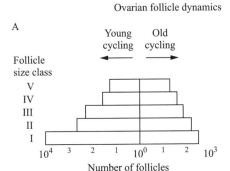

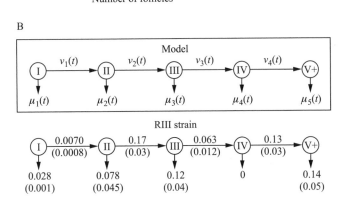

Figure 2.2. Ovarian follicles size classes. Redrawn from Faddy et al. (1983). *A*, The growing follicle populations can be represented as a pyramid of ascending size classes of growing follicles, which include progressively fewer of each larger size. Size class I is primordial follicles, which have a single layer of granulosa cells and are not active in steroid production (Fig. 2.1). As follicles mature (classes II–V), and under the influence of FSH (follicle stimulating hormone) the granulosa cells proliferate and secrete increasing amounts of estrogens. Class V is antral follicles, some of which will burst at proestrus (ovulation) to shed their mature ovum for possible fertilization. As the ovary ages, and the primordial follicles are irreversibly lost, the follicular pyramid becomes increasingly steeper, because the growing size classes are less affected. *B*, For each stage in follicle maturation, rates can be calculated at which a follicle and its oocyte will transit (migrate) to the next stage $v(t)$ or die at that stage $\mu(t)$.

(hemi-ovariectomy or hemispaying), the remaining ovary almost immediately compensates by increasing the production of normal ova that can be fertilized. The compensatory effect of hemispaying has been obtained in lab rodents, livestock, and other mammals. This experiment was first reported by the remarkable John Hunter in 1777, who showed that the hemispayed sow produced nearly as many (87%) young as the intact sow during the next eight litters (Short, 1977). The production of ova by the remaining ovary doubled as the result of increased secretion of pituitary hormones. Clearly, at least some of those follicles and ova that would have been lost in the remaining ovary are viable.

Individual ovaries of neonates vary at least twofold in the numbers of primordial follicles and oocytes in both humans and mice (see Fig. 1.3). Because of the ongoing death of follicles, the numbers diminish strikingly during aging. The rate of loss follows exponential kinetics, in which a fixed fraction of oocytes and follicles are lost per unit of time. Thus, the probability of oocyte loss is independent of the total number remaining. These kinetics resemble the decay of radioactivity and are fitted by a linear regression of log(number total follicles or oocytes) versus age from birth for most of the reproductive life span (Jones and Krohn, 1961a,b; Nelson and Felicio, 1986; Gosden and Faddy, 1998; Fig. 2.3).

However, at the very end, when >95% of oocytes have been lost, the rate accelerates in humans (Richardson et al., 1987; Fig. 2.3). Similarly, in young mice whose ovaries were partially depleted of oocytes by treatment with busulfan, an alkylating agent used in chemotherapy, there is accelerated entry of primary follicles into the growing pool (Hirshfield, 1994). This accelerated loss could be a consequence of increased secretions of pituitary gonadotrophin that are a feedback response to dwindling follicle reserves. Mice that were genetically engineered to sustain high blood levels of the pituitary LH had a premature loss of primary follicles (Flaws et al., 1997). These findings show complex interactions between the ovary and neuroendocrine controls in determining when the ovary is depleted of oocytes.

Individuals differ widely in the numbers of remaining occytes as ovarian depletion advances. By middle age at the onset of reproductive senescence in mice (10–15 months), oocyte numbers vary from none in exhausted ovaries of acyclic individuals up to 1000 in others of the same age that are able to maintain regular fertility (estrous) cycles (Gosden et al., 1983; Fig. 2.4A). These studies were done in the lab of C.E.F. in collaboration with Roger Gosden and with James Nelson and Leda Felicio (the latter were graduate students at USC at that time). Subsequent studies of premenopausal women by J.N. showed a 1000-fold range of oocytes (Fig. 2.4B; Richardson et al., 1987). The cycle length also tends to vary increasingly widely before acyclicity in both mice and humans (Fig. 1.2C,D). Some degree of increasing variability is predicted from stochastic models of ovarian follicle survival (Fig. 2.2), even in the hypothetical case that all individuals begin life with equal numbers of follicles and oocytes.

In laboratory rodents, individual ovulatory cycles are easily followed by vaginal smears (section 1.3). Inbred mice of the same birth cohort, which were tracked by daily vaginal lavages during their reproductive life spans, showed individual patterns of fertility cycle frequency and changes during aging. Some mice lost regular fertility cycles by 9 months, whereas cycles persisted in others to 18 months or later (Fig. 1.3). Mice with fewer oocytes in their ovaries would be expected to lose fertility cycles earlier.

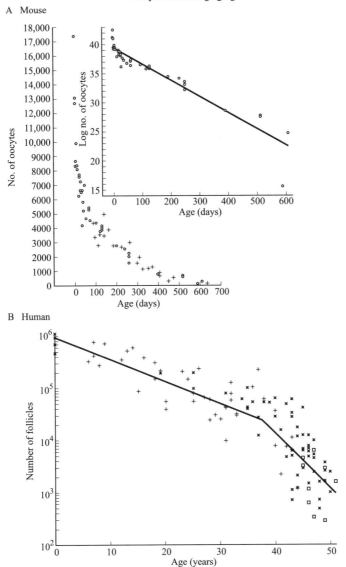

Figure 2.3. The rate of ovarian aging, measured by the loss of primary follicles, each of which contains a single oocyte. *A*, RIII mice. The probability of oocyte loss is independent of the total number remaining. These kinetics resemble the decay of radioactivity and fit a linear regression of log(number total follicles or oocytes) versus age. Redrawn from Jones and Krohn (1961a). *B*, Humans, [Richardson et al. (1988)] including neonates from Block (1952, 1953, also shown in Fig. 1.3). Note the accelerating loss of oocytes after >95% of oocytes are lost. From Faddy and Gosden (1992) redrawn in vom Saal et al. (1994), with permission.

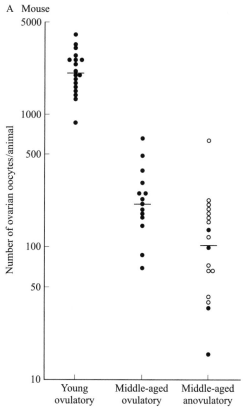

A Mouse

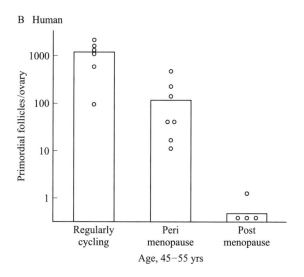

B Human

However, age changes in the neuroendocrine centers in the brain are also involved in reproductive aging. Species differ in the relative role of age changes in the ovary and the hypothalamus to reproductive senescence. In humans, the strongest influence on menopause may be ovarian oocyte depletion, whereas rodent reproductive senescence also involves hypothalamic changes that interact with ovarian steroids (Finch et al., 1984; vom Saal et al., 1994; Wise et al., 1996). Here, we focus on the ovary.

Humans also have individually characteristic cycle frequencies (Fig. 1.2B). Even identical twins vary in the age of menopause almost as much as their nonidentical sisters (section 1.3; Fig. 1.2E). It is not known whether these differences are due to neuroendocrine output of gonadotropins, or to responsiveness of ovarian follicles to hormones, or both. As menopause approaches, cycles become more irregular (Treloar et al., 1967; vom Saal et al., 1994). The most comprehensive study of oocyte numbers in peri-menopausal women indicates imminent depletion of the remaining follicles and oocytes (Richardson et al., 1987; Fig. 2.4B). However, these data were based on surgical specimens. It is hard to obtain data on oocyte numbers in healthy women, because ovaries are mainly obtained from surgery of women being treated for tumors or other abnormalities in their reproductive tracts, which could bias the findings. The individual variations in the age of onset and the particular fluctuations in cycle length during aging of inbred mice approximate those of general human populations. Thus, a vital parameter of life history, the age of menopause in humans, appears subject to considerable nongenetic variability. Next we consider how variations in oocytes influence reproductive senescence.

2.2. Oocyte numbers determine menopause

The number of surviving ovarian oocytes by middle age is a major determinant in reproductive senescence in female rodents. As shown above, inbred mice differ > 1000-fold at middle age in the numbers of surviving oocytes (Fig. 2.4A). These and other data indicate that at least 150 follicles and oocytes are required for mice to maintain regular fertility

Figure 2.4. The capacity for fertility cycles depends on the numbers of remaining ovarian oocytes and is not strongly predicted by chronological age. During middle age in female mice and humans, individuals differ widely in the remaining numbers of oocytes. A, By middle age, at the onset of reproductive senescence in female mice (C57BL/6J; 10–15 months), oocyte numbers vary from 0 in exhausted ovaries in a prematurely senescing mouse to > 1000 in mice of the same age with regular fertility (estrous) cycles. From Gosden et al. (1983), with permission. B, Premenopausal women (45–50 years) also vary > 1000-fold in the numbers of remaining oocytes and follicles. Redrawn from Richardson et al. (1987).

cycles; below this threshold number, regular cycles cease (Fig. 2.4A; Gosden et al., 1983; Nelson and Felicio, 1986). For humans, a threshold number of about 1100 follicles is required to maintain regular cycles (Fig. 2.4B: Faddy and Gosden, 1995, 1996; Gosden and Faddy, 1998). In modeling human menopause, this mean value of 1100 follicles was varied randomly (Faddy and Gosden, 1995, 1996; Gosden and Faddy, 1998); the theoretical distribution of ages at menopause closely matched that observed in a longitudinal study (Fig. 2.5; Treloar, 1981). The number of fertile years equals 0.077 times the square root of the numbers of oocytes that remain, for individuals with < 100,000 follicles (Gosden and Faddy, 1998).

Because oocytes are not replaceable, the *initial* stock of follicles and oocytes should determine the numbers surviving to middle age. This hypothesis was tested by a powerful experiment, in which Nelson and Felicio (1986) evaluated how the numbers of oocytes present in the young ovary influence the time of reproductive senescence. The reserve oocyte pool was diminished in young adults by subtotal ovariectomy. Assuming that at least 150 follicles are needed maintain cycling, Nelson and Felicio

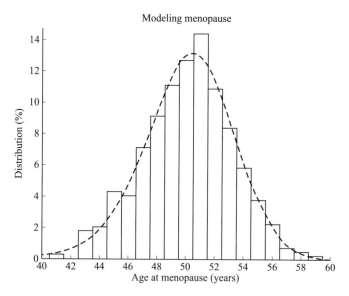

Figure 2.5. A stochastic model for age at menopause, based on threshold number of oocytes for maintaining regular fertility cycles (Faddy and Gosden, 1996). Gosden and Faddy (1998) estimated for humans that about 1100 follicles are the threshold number of remaining follicles required to maintain fertility cycles. In a modeling study, this mean value was programmed to vary stochastically in a hypothetical population of women and gave a distribution of ages at menopause that closely matched the data of Treloar (1981) from longitudinal self-reports on cycle frequency up through menopause (Fig. 1.2B and D). Redrawn from Gosden and Faddy (1998).

calculated from the regression equation of total follicles versus age that a 90% surgical reduction of ovarian oocytes and follicles should accelerate reproductive senescence by 7 months. Readers may note that the age of reproductive senescence is not *directly* proportional to the number of follicles and oocytes, because the ovarian oocyte depletion declines exponentially. The results from longitudinal daily observations of vaginal cytology confirmed this prediction for the age-related lengthening of cycles as well as the onset of acyclicity (Fig. 2.6). This experiment also shows the limits of neuroendocrine compensation to the loss of oocytes.

These findings support the hypothesis that the variable numbers of oocytes as determined before birth are *the* key factor in the age of natural menopause, which occurs five decades later in humans. However, neuroendocrine mechanisms may also be important in regulating the decline, which accelerates in association with elevated gonadotropins as noted above (Fig. 2.3B). Although the simplest hypothesis is that the remaining ovarian mass in premenopausal women determines the gonadotropin output, there could be separate hypothalamic age changes (Wise et al., 1996). Because gonadotropin levels govern the rates of follicular atresia, which, for example, are slowed by the absence of gonadotropins after hypophysectomy (Jones and Krohn, 1961b), we do not exclude a

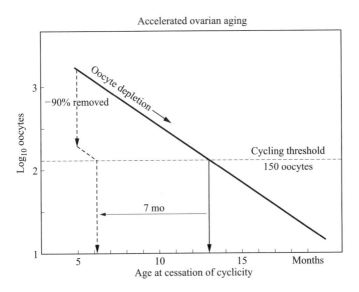

Figure 2.6. Model for ovarian aging. Assuming that at least 150 follicles are needed to maintain cycling in female mice, Nelson and Felicio (1986) calculated from the regression equation of total follicles versus age that a 90% surgical reduction of follicles should accelerate reproductive senescence by 7 months. The results from longitudinal daily observations of vaginal cytology confirmed this prediction for both the age-related lengthening of cycles and the onset of acyclicity. Redrawn from Nelson and Felicio (1986).

role of neuroendocrine modulation in the timing of menopause (Finch et al., 1984; Wise et al., 1996).

We must also acknowledge species differences in aging primates. Rhesus monkeys are very like humans in their loss of fertility and ovarian oocyte depletion with ovariprival hot flushes, which occur before the last 20% of their life span. However, chimpanzees retain fertility until just before their life span limit; no data are available on oocyte or follicle numbers (reviewed in Finch and Sapolsky, 1999). These species differences imply considerable evolutionary flexibility in ovary–neuroendocrine interactions that allow wide species variations in the schedule of fertility loss during aging. Next, we consider how these oocyte variations arise before birth.

2.3. Development and ovarian oocyte variations

Twofold individual variations of primary follicle numbers are present even before birth (Block, 1953; Baker, 1963). Chance operates here on two processes during the first third of gestation during the formation of the ovary: (1) loss of germ cells during migration and (2) selection of cell fates. In embryos of all vertebrates and in many invertebrates, the future oocytes can be traced as clonal lineages of promordial germ cells (PGCs), which undergo extensive proliferation and migration. In many animals, PGCs can be identified early in development by characteristic cytoplasmic granule markers, which were available long before the molecular era provided protein or RNA markers (Davidson, 1986; Lin and Spradling, 1993; Buehr, 1997). In mammals, as well as most other animals, the PGCs differentiate during gastrulation in the extraembryonic mesoderm. They then migrate from there to the site of the future ovary or testes (Tam and Snow, 1981; McClaren, 1992; Buehr, 1997). The signals that guide these cell movements include chemotactic proteins in the extracellular environment that guide migratory PGCs (for example, TGFβ, Steel factor) and integrin receptors on PGCs that modify cell adherence to extracellular matrix glycoproteins (laminin, fibronectin, collagen IV); cell–cell interactions between the PGCs are also prominent (Garcia-Castro et al., 1997). The particular signaling pathways do not concern us here.

During the mouse gestation period of 20 days, PGCs originate in the extraembryonic mesoderm at day 7, as a founder population of about 100 cells (Buehr, 1997; Hirshfield, 1997; McLaren, 1997). This strange origin of PGCs outside of the embryo itself is speculated to protect PGCs from gene imprinting by DNA methylation ongoing in the embryo proper at this time. During the next 4 days when the gonads and other organs are formed, PGCs migrate on a complex path that takes them into the embryo and through at least four different tissue beds (Tam and Snow, 1981; McClaren, 1992; Buehr, 1997). During this roundabout march, the PGCs continue to proliferate so that many thousands eventually arrive in the ovary. After leaving the base of the allantois outside of the forming embryo, the PGCs penetrate through the wall of the gut (day 8.5), then move along within the dorsal mesentery (day 9.5), and finally reach their ultimate destination, the genital ridges (days 10.5–11.5), where they collect at the site of the future ovary or testis. During this migration, adhesivity of PGCs to extracellular glycoproteins decreases, particularly for laminin, which is present at high concentrations in the germinal ridge (Garcia-Castro et al., 1997). Garcia-Castro et al. (1997) postulate cooperative interactions between PGCs and laminin at the nascent gonad, which would generate chance outcomes in the duration of contact, thereby contributing to the heterogeneity in subsequent proliferation and survival discussed below.

Up to this stage, male and female embryos are indistinguishable. At day 13.5, PGC proliferation ceases and meiosis begins, although the transition from mitosis to meiosis is not synchronous. What governs these statistics is not known. The oocytes remain in arrested meiosis (prophase I) until full reproductive cycles begin after puberty. Shortly after birth, PGCs gradually become surrounded by follicle cells to form the primordial ovarian follicles (Hirshfield and DeSanti, 1995). Twofold individual variations in the numbers of PGCs are observed throughout these early stages. Using published data on individual mice from the same litter of inbred parents (Tam and Snow, 1981), we calculated that C-VAR of 23–32% across all stages (prenatal days 8.5–13.5; Fig. 2.7A). The variability in initial numbers of oocytes (32% C-VAR) approximates that in young adult ovaries (Table 1.4). The ovary may be the first organ for which the profile of chance variations in cell numbers has been computed from early development into adult life. Because inbreeding often increases variability (see Note, p. 75), these statistics need to be verified in human ovaries.

Three factors contribute to the variable numbers of PGCs and oocytes in the embryonic ovary: variable cell proliferation, variable cell death, and navigation errors during cell migration. The first factor, variable proliferation, appears to be the major cause, as indicated by PGC proliferation in primary cell cultures, which cloned single PGCs from 8.5 day donors (Ohkubo et al., 1996). The rate of proliferation was initially very high in these clones, but then declined sharply on the same schedule in vitro as in the normal embryo. The individual PGCs varied widely in the numbers of divisions, as evaluated by PGC number per colony (Fig. 2.7B). The largest size classes were colonies with one to three PGCs, indicating that most of the PGCs from day 8.5 embryos divided only once. However, a minority of other PGCs divided four or more times, to give colonies of up to 48 PGCs. This pattern of PGC growth arrest indicates a random process of cell fate determination, which prevents further proliferation and which is cumulative over time, eventually rendering all PGCs postmitotic. Many other cell types display similar heterogeneity in proliferative potential, which can be described in terms of stochastic processes (chapter 3). Mixtures of PGCs from

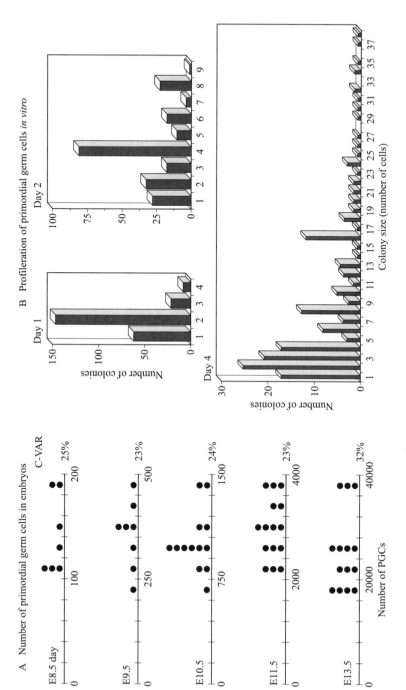

Figure 2.7

embryos of different ages did not alter the colony size, which indicates few cell–cell interactions and that the shutoff of proliferation was programmed early during PGC differentiation. In humans, these processes generate millions of PGCs prenatally; the statistics of this expanded proliferation are not known but lead to a fourfold range of oocytes at birth (see Fig. 1.3).

Various growth factor proteins modify the extent of cell proliferation and cell survival; for example, LIF (leukemia inhibitory factor) and SCF (stem cell factor) block PGC death in vitro (Pesce et al., 1993; Buehr, 1997). However, the subcellular mechanisms in the apparently random shutoff of proliferation are not known. In rare embryos, a few PGCs continue to proliferate and become teratocarcinomas that can migrate throughout the embryo and become tumors in the embryo or later. Cell variations in the density of receptors for growth factors could also influence the extent of proliferation.

The contribution of cell death to variations in the initial stock of PGCs is unclear. PGC death during migration and in the early embryo has many characteristics of programmed cell death, or apoptosis (Pesce et al., 1993; Coucouvanis et al., 1993, 1995; Buehr, 1997). The first major phase of apoptosis occurs in oogonia at day 13 in the mouse embryo, which is after most PGCs ceased proliferating; a second phase strikes oocytes at the pachytene stage of meiosis between day 15 and birth. Death of PGCs can be observed in primary PGC cell cultures from the embryo (Fig. 2.7B), which suggests that up to 50% of the PGCs die between day 4 and day 5 in culture (Ohkubo et al., 1996). Remnants of dead PGCs were seen later during culture, at a time when individual colony size had decreased. The size classes of colonies with odd numbers of PGCs could arise from cellular heterogeneity in the extent of proliferation, as well as from cell death. The mechanisms that determine sur-

Figure 2.7. Primordial germ cell (PGC) numbers in the early mouse embryo. During the 20-day gestation, PGCs are first seen in the extraembryonic mesoderm at day 7 as a group of 10–100 cells. During the next 4 days when organs are formed, PGCs proliferate during their migration into the embryo. The PGCs migrate in columns through the wall of the hindgut (day 8.5), then within the dorsal mesentery (day 9.5), and finally to the genital ridges, where they collect at the site of the future ovary (days 10.5–11.5). By day 13.5, the ovaries have their maximum numbers of oocytes. *A*, The frequency distribution of PGC numbers per embryo (redrawn from Fig. 7 of Tam and Snow, 1981), with *C-VAR* calculated from their table 3. *B*, Clonal studies of PGC proliferation during primary culture. Single PGCs were plated and colony size (numbers of PGCs) counted during the following 5 days, which corresponds to the end of PGC proliferation and ovary formation in vivo. The 50-fold range of colony size shows clonal variations in proliferation and in cell death. Redrawn from Ohkubo et al. (1996).

vival or death are unknown but may involve some of the same growth factors that determine PGC proliferation (see above). A working hypothesis is that PGC death is caused by insufficient local concentrations of a required growth factor (Coucouvanis et al., 1993).

Navigational errors during PGC migration do not appear to be important in later oocyte variations, because only a few (1–2%) PGCs fail to reach the ovary in day 13.5 mouse embryos (Tam and Snow, 1981). Occasional strayed (ectopic) PGCs are found in neonates, for example, adrenals of mice (Upadhyay and Zamboni, 1982; Buehr, 1997) and in the body wall of neonatal fish (Kobayashi and Hishada, 1992). The failure of ectopic oocytes to survive into adulthood indicates that they gradually die off. The time and place of death of ectopic PGCs during development is not known.

The maximum number of PGCs is reached before birth, with totals for both ovaries of about 25,000 in mice and 2 million in humans (Baker, 1963). Once the initial stock of follicles has formed, large numbers of them die (Fig. 2.3). In humans and primates, about 70% of the initial stock of oocytes die before birth (Rabinovici and Jaffe, 1990). Follicular loss through cell death (atresia) continues on a massive scale during postnatal development and into adult life. We do not know if the same mechanisms that cause death of PGCs during formation of the ovary also cause follicular atresia in the embryo ovary once it has formed, or in the postnatal ovary. Further studies are needed to resolve the sources of variation. Postnatal follicle atresia is greatly suppressed in mice that are *Bax*-deficient (Perez et al., 1999). Remarkably, *Bax*-deficient mice retain large stores of primary ovarian follicles at ages approaching the life span. Manipulations of Bax activities, a protein which mediates apoptotic cell death in many organs, might give an approach to slowing the loss of oocytes during aging.

We briefly consider oogenesis during aging in other species. In possible contrast to the general mammalian pattern (see Table 1.1), there are indications of de novo oogenesis in adult dogs and several prosimians (reviewed in Finch and Sapolsky, 1999; Kumar, 1974), but there is no demonstration that the putative germ cells become viable ova. The extent of PGC proliferation in adult fish, amphibians, and reptiles is not known but may be much greater than in mammals because of the persistence of fertility to advanced ages in the few very old individuals examined (Finch, 1990, 1998). For example, rockfish (*Sebastes*) from the North Pacific live more than 100 years without indications of reproductive senescence. A few aged specimens examined showed histological evidence of seasonal oocyte maturation (deBruin, Gosden, Finch, and Leaman, unpublished ms.). Fruitflies, like most other animals, share similar patterns of ovary formation, with early differentiation of PGCs and extensive migration; ectopic PGCs die rapidly (Poirie et al., 1995). Adult fruitflies also show great variation among individuals within strains; Robertson (1975a) reports *C-VAR* of 11–15%. However, fruitflies do not show marked decrease in egg production during aging, possibly because PGCs continue to proliferate in adults (see Table 1.1). In *Caenorhabditis*, egg laying decreases markedly, with a prolonged sterility phase before death that may be up to 50% of the life span; however, little is known about oogenesis at later ages (Table 1.1).

To summarize, mammals are subject to two main sources of individual differences in oocyte numbers even before birth: chance variations in the numbers of PGCs that proliferate during development, and chance outcomes in the numbers of follicles lost during atresia. Even if genetically identical individuals began adult life with identical numbers of oocytes and follicles, the chance processes that lead some follicles to die and others to progress further in maturation would, over time, led to major individual variations in the numbers of follicles at midlife, and hence variations in the age of menopause. Provisionally, we suggest that most of the individual variation arises from stochastic differences in the PGC proliferation, whereas errors in PGC migration appear to be relatively small. Other mammalian cells also show stochastic features in proliferative capacity (section 4.2). The variance of primary follicle numbers warrants further characterization in the postnatal ovary to extend our knowledge beyond rodents. For example, it would be most illuminating to have some ovarian histologic data on human twins.

2.4. Maternal age factors in birth defects

Birth defects from abnormal numbers of chromosomes (aneuploidy) may also be the result of chance events. The risk of fetal aneuploidy increases exponentially after 30 years of age in humans (Fig. 2.8). In humans, Down syndrome (trisomy 21) accounts for the majority of live-born aneuploidy. In >85% of cases, the nondisjunction is attributed to maternal origins that arose before fertilization. The maternal age risk doubles with every 6 years after 30, so by the maternal age of 50 years nearly 10% of fetuses born are aneuploid (Hook, 1981; Gosden, 1985). Rodents also show maternal age effects on fetal aneuploidy (there are genotypic differences) (Fabricant and Parkening, 1978; Eichenlaub-Ritter et al., 1988; Finch, 1990).

Down syndrome adults suffer a further insult, premature dementia, with onset of the neurodegenerative changes of Alzheimer disease (Schupfer et al., 1998; Hardy, 1997; section 5.1.3). The mechanism may involve the extra copy of the β-amyloid precursor protein gene (β-APP), which is located on chromosome 21 and which gives rise to the Aβ peptide found in senile plaques. Because Down syndrome life spans are increasing toward the normal range through improved care and increased social and employment opportunities, increasing numbers are living long enough to experience the further disability of early-onset Alzheimer disease.

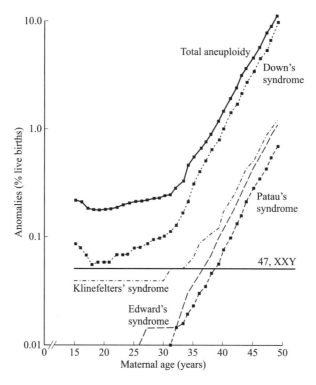

Figure 2.8. Maternal age effects on fetal aneuploidy, in which trisomy 21 (Down syndrome) accounts for the majority of defects. Other aneuploid conditions that follow the same schedule (in order of frequency) are the syndromes of Klinefelter (XXY; leading to sterility and breast development), Edward (trisomy 18), and Patau (trisomy 13). Redrawn from Gosden (1985).

The size of the oocyte–follicle pool, not age per se, may be a major factor in maternal age risk of fetal aneuploidy. An elegant experiment by Gosden and colleagues showed that removal of one ovary increased the incidence of fetal aneuploidy twofold in mice (Brook et al., 1984) and also increased abnormal chromosome alignments (Eichenlaub-Ritter et al., 1988).

These findings indicate that the main maternal factor in fetal aneuploidy is the diminishing pool of ovarian oocytes, rather than chronological age. A similar situation occurs in women with Turner syndrome. This rare aneuploidy (45, X) causes gonadal dysgenesis and greatly reduces the initial pool of oocytes, so menopause generally occurs prematurely before age 30 (E. Magensis, pers. commn., cited by Brook et al., 1984). In parallel with hemi-ovariectomized mice, during their brief reproductive years the 1–2% of Turner women who, through mosaicism, are fertile have a greatly increased risk of having Down syndrome offspring (Brook et al., 1984; Tarani et al., 1998).

These findings in mice and Turner syndrome humans further implicate the chance events during development that determine the size of the initial oocyte pool, and indirectly at the end of reproductive phase, the risk of fetal aneuploidy.

It is not known why aneuploidy is increased as the pool of oocytes diminishes. One long-standing hypothesis is that defective oocytes are eliminated during the ongoing loss of oocytes, which begins in the ovary even before birth. The primary oocyte remains in a prolonged state of meiotic arrest, at prophase I (dictyate, or diplotene stage), after it has formed during development. Meiosis is resumed in the mature ovarian follicle under the control of gonadotropins just before ovulation, which may happen 50 years later in humans, elephants, and whales. During this long sojourn, the mitotic spindle apparatus may be subject to chance molecular damage or disorganization. The greater sensitivity of oocytes in middle-aged mice to arrest by the antimitotic agent colchicine at metaphase I (after reactivation from dictyate) suggests increased instability in spindle functions (Tease and Fisher, 1986). Because spindle microtubule proteins undergo continuous cycles of de- and repolymerization, there are many opportunities for chance errors in spindle organization and attachment to chromosomes that may lead to fetal aneuploidy (Nicklas, 1997). Random oxidative damage could also be a factor

A decrease of chromosomal recombination with maternal age is widely observed. The basic rule of genetic exchange is that the frequency varies in proportion to the distance separating the loci on a chromosome, another deal by chance. During early aging, fruit flies show a 90% decrease in crossover frequency (Bridges, 1927). However, mice show milder and more progressive decreases in females, but not males (Henderson and Edwards, 1968; Harman and Talbert, 1985; Finch, 1990, p. 452–454). Henderson and Edwards (1968) proposed that the decreased chiasma formation contributes to fetal aneuploidy because of the inverse relation between chiasma formation and nondisjunction leading to univalent (unpaired chromosome).

Last, we suggest that, if fetal aneuploidy can be increased by hemiovariectomy, then it may eventually be possible to reduce aneuploidy through interventions that slow the rate of oocyte loss. Two interventions that slow the rate of oocyte loss in adult mice are removal of the pituitary (hypophysectomy; Jones and Krohn, 1961b) and food restriction (Nelson et al., 1985). There are no clear indications that suppression of ovulation by steroidal contraceptives or anorexia slows oocyte loss or delays menopause. As mentioned in section 2.2, the loss of ovarian oocytes might also be slowed by synthetic peptides that antagonize actions of LH. A larger oocyte pool might thereby reduce the risk of fetal aneuploidy at later maternal ages. Next we consider how variations in the age of menopause may influence outcomes of individual aging.

2.5. Health risks associated with menopause

Here we consider diseases that ensue after menopause. In particular, cardiovascular disease and osteoporosis have estrogen-sensitive risk factors, which may be indirectly influenced by the initial oocyte pool size. In many industrialized countries, men have an earlier increase in the risk of cardiovascular disease that begins before 40, whereas marked increases for women are not seen until after menopause. For example, age-standardized mortality from cardiovascular disease in women is about 15% that of men aged 35–44 years (Uemura and Pisa, 1985). Based on the experiment showing that mice with fewer initial numbers of oocytes and follicles have earlier reproductive senescence (Fig. 2.6), we hypothesize that a woman's initial numbers of ovarian oocytes and follicles determine the risk schedule for cardiovascular disease and osteoporosis. For evidence, we consider the positive effects of estrogen replacement therapy (ERT) on heart disease, osteoporosis, and Alzheimer disease, and the effects of premature menopause.

The risk of heart attacks increases after menopause in association with increased blood cholesterol and higher ratio of low- to high-density blood lipoproteins (LDL:HDL). More than 30 studies agree that ERT after menopause or surgical removal of the ovaries reduces the risk of heart attacks by about 50% and improves lipid and vascular endothelial risk factors (PEPI Trial Writing Group, 1995; Stampfer et al., 1991; Barrett-Conner and Bush, 1991; Beale and Collins, 1996; Lip et al., 1997; Nasr and Breckwoldt, 1998). Although a few studies have shown no effect or negative effects of ERT (cited in Beale and Collins, 1996; Nasr and Breckwoldt, 1998), the general trend for improved cardiovascular health is now widely recognized. The mechanisms of ERT include reversal of the shift in LDL:HDL lipids, which may account for less than 50% of the benefits. However, estrogen replacement has numerous other positive indirect effects, including increased growth hormone and decreased platelet and coagulation activities, as well as direct activities as an anti-oxidant (the latter is shown in experimental models, and its contribution in vivo is unknown). Ongoing studies are evaluating ERT for more than 10 distinct biochemical processes that promote atherogenesis (Beale and Collins, 1996; Nasr and Breckwoldt, 1998), including two major double-blind randomized trials, the Woman's Health Initiative and the Heart Estrogen-Progestin Replacement Study. Other studies are examining combinations of steroids to reduce the risk of breast cancer currently associated with ERT (Henderson et al., 1993).

Women begin a slow loss of compact bone (osteoporosis) during the fourth or fifth decades, which accelerates after menopause, about 20 years earlier than men, as shown in Figure 2.9A. These graphs also show the wide range of individual variations in bone density. In women, studies of twins and mother–daughter pairs indicate that heritable factors account for 50–75% of bone density variance (Arden et al.,

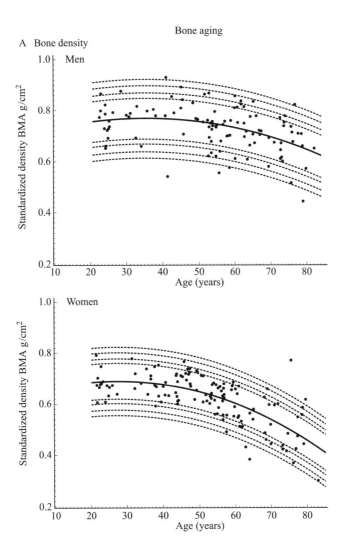

Figure 2.9. Bone density and aging. *A*, Bone density was measured in
trabecular bone of the forearm (radius) in men (*top*) and women (*bottom*).
Eight European centers participated in the European Union Concerted Action
for Quantitative Assessment of Osteoporosis, which has developed a highly
reliable cross-calibration in bone densitometry, to normalize for different
instrumentation across centers. The radius bone was selected because of its risk
in Colles's fracture (see B, next page). Women had lower bone density at all
ages than men, with greater differences emerging after menopause. The scatter
plots for individual values of standardized density also show contour lines for
70%, 80%, 90%, and 95% confidence intervals. From Reeve et al. (1996), with
permission. *B*, Women show early increase in risk of fractures during aging
after menopause, although men eventually reach the same levels of risk. From
a predominantly European-origin population in Rochester, Minnesota.
Redrawn from Riggs and Melton (1986).

1996; Nguyen et al., 1998; Ferrari et al., 1998a,b). However, bone density at any age is determined by many other factors besides estrogen (physical activity, smoking, nutrition, growth hormone and other hormones). The sharp decrease of estrogens at menopause greatly accelerates the risk of osteoporotic fractures (Fig. 2.9B). These processes are modeled in Figure 1.9, as accelerated aging. Radiological studies indicate a threshold level of about $0.8\,\mathrm{g/cm^2}$ bone density for high risk of spontaneous fractures (Ito et al., 1997; Resch et al., 1995). (This is another case of the general threshold model for effects of aging; see Figs. 1.9 and 2.6.) Osteoporotic fractures, in turn, increase mortality and morbidity (Cauley et al., 1997; Cooper, 1997; Diamond et al., 1997).

Further evidence for the role of estrogen deficits in osteoporosis versus other age changes is that premature menopause triggers earlier osteo-porosis. Japanese who became menopausal before 43 years had 20% less bone density in lumbar vertebrae than those who were 11 years older and who more recently had become menopausal (Ohta et al., 1996). Among the prematurely menopausal, 55% were below the threshold of bone density for high risk of fracture, whereas < 10% of the older, more recently menopausal women had this extent of bone loss. The role of chance in osteoporotic fractures may thus include variations in the ovarian folicular pool. However, the nonheritable portion of indi-vidual variance in bone density (25–50%; see above) could arise from

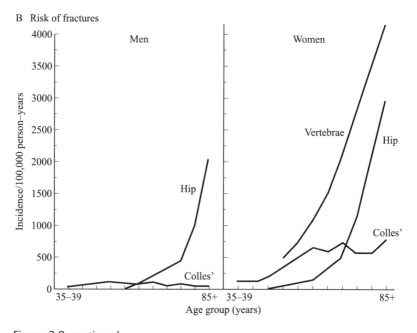

Figure 2.9 continued

chance variations in bone structure and its numbers of bone-forming cells, as well as lifestyle choices of diet and exercise.

Ovarian follicle reserve is also a factor in a side effect on the ovary of cancer treatment. Chemotherapeutic alkylating agents like cyclophosphamide cause premature menopause in older women at lower doses than in younger women (Headley et al., 1998; Bines et al., 1996). In parallel with this age sensitivity, ovarian follicles of congenitally hypogonadal mice are more sensitive to radiation (Gosden et al., 1997). Thus, women with smaller ovarian reserves at any age would be more sensitive to alkylating agents. Women who became perimenopausal from ovarian damage during chemotherapy for breast cancer (Headley et al., 1998) or lymphoma (Ratcliffe et al., 1992) had lower bone density than those receiving chemotherapy who continued to have menstrual cycles.

Alzheimer disease also appears to interact with postmenopausal estrogen deficits. From several post hoc studies, ERT users had lower incidence of Alzheimer disease (Pagannini-Hill and Henderson, 1994; Schneider and Finch, 1997; Tang et al., 1996; Kawas et al., 1997; Shadlen and Larson, 1999), but findings are mixed (Yaffe et al., 1998; Haskell et al., 1997). Pilot data indicate that ERT may enhance responses to cholinergic medications used in Alzheimer disease (Schneider et al., 1996). A precedent for ERT efficacy is the response of younger women after surgically induced menopause, whose rapid onset of impairments of cognitive functions was prevented by ERT (Sherwin, 1998; Sherwin and Tulandi, 1996). However, the abrupt declines of ovarian steroids after ovariectomy may have different effects on brain functions than those during menopause, which are spread over several years. Ongoing studies of cognitively tested women may resolve effects of ERT on cognitive age changes that are distinct from specific pathological conditions.

In sum, because the onset of the loss of ovarian estrogen production at menopause is tightly linked to the depletion of estrogen-producing follicles, we hypothesize that individual variations in the age of menopause and subsequent estrogen-deficit-dependent disorders can in part be traced back to developmental variations in the numbers of follicles present at birth. Nonetheless, we must reckon with the extraordinary 122 year life span of Jeanne Calment, whose postmenopausal phase of about 70 years was twice that of her early adult years with active ovaries. Mme. Calment had menopause about 20 years before ERT became available in a few clinics. She never used ERT and enjoyed good physical and mental health for most of those 70 years (Allard et al., 1998; Ritchie, 1995). We have no idea why some women have few health consequences of menopause for many decades without ERT, which may involve some combination of inherited genes, lifestyle, and developmental variations.

2.6. Fetal steroid variations and reproductive system development

In this section we consider information on variations in fetal sex steroid levels of mammals, which are surprisingly large and which are candidates for causing quantitative variations in sex-steroid-sensitive cells in the brain and in the prostate.

2.6.1. Brain

Here we give an overview of fetal gonadal determinants of reproductive system development, which by its brevity cannot acknowledge all the main contributors to this remarkable literature (Byne, 1998; Jost, 1965; George and Wilson, 1994; Resko and Roselli, 1997; McEwen, 1997; Swerdloff et al., 1992). Soon after the ovary has formed in the human female fetus at 2 months, some of follicles have begun to grow and produce steroids, which increases fetal blood estradiol for the remainder of gestation (Fig. 2.10). Similarly, in the male fetus, the Leydig cells secrete testosterone by 2 months (Fig. 2.10). Testosterone synthesis is maximum at midpregnancy and declines thereafter in males. The gender differences in sex steroid production are due to different schedules in the activation of genes that encode enzymes of steroid synthesis (Greco and Payne, 1994). Elevations of fetal androgens, in turn, influence development elsewhere in the reproductive system. Testosterone is a prohormone in both sexes and is converted to estradiol by the enzyme aromatase in brain and other organs. Both estrogens and androgens have roles in male and female development, and both are present to some degree in fetal blood. Female reproductive tract development does not require the presence of the ovary. Removal or congenital deficiency of the fetal testes causes the reproductive tract in chromosomal males to become female in morphology. Thus, a default program for gender development directs female development. Besides the role of fetal estrogens and androgens in brain development, progesterone of maternal origins is implicated (Wagner et al., 1998). The details of these sex steroid programming processes involve both masculinization and defeminization as mediated by genes on both sex chromosomes (Jiminiz and Burgos, 1998; Swain et al., 1998; Hagg and Donahoe, 1998; see Fig. 2.11 legend), but to tell this story would extend beyond our focus on aging.

Many organ systems that differ between the sexes have critical periods for their development, after which they acquire their adult destiny, although adult levels of hormones may be required for those functions (Fig. 2.11). Sex differences arise during development in many brain regions, such as the corpus callosum and hypothalamus; the latter contains prominent sexually dimorphic nuclei (SDN; Figs. 1.8, 2.12). In rodent models, manipulations of fetal and neonatal sex steroids modify neuron numbers in the SDN (Dodson and Gorski, 1993; Zhou et al.,

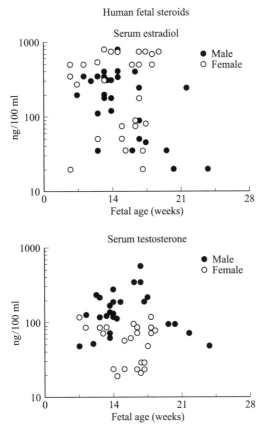

Figure 2.10. Human fetal serum sex steroids in the first trimester. These major 1000% (10-fold) variations among individuals far exceed the variations of the assay procedure (9% *C-VAR* from pooled serum that represents the high and low range). *Top*, Estradiol. *Bottom*, Testosterone. Redrawn from Reyes et al. (1974).

1995; Resko and Roselli, 1997). Species differ widely in the schedule of these events. In rodents, some features of brain function continue to be programmed by sex steroid levels during the first postnatal week, whereas in primates the postnatal sex steroid levels appear to be less important to adult brain characteristics (Clark et al., 1989; Goy et al., 1988). Species also differ in the extent to which the chromosomal sex influences hypothalamic–pituitary controls. For example, male monkeys that received ovary transplants had blood estrogen indicative of growing follicles, which even ovulated and formed a corpus luteum (Norman and Spies, 1986). The corresponding ovary transplant in male rats does not function, nor can adult male rats respond to estrogen elevations with a preovulatory surge of gonadotropins. These species differences give cau-

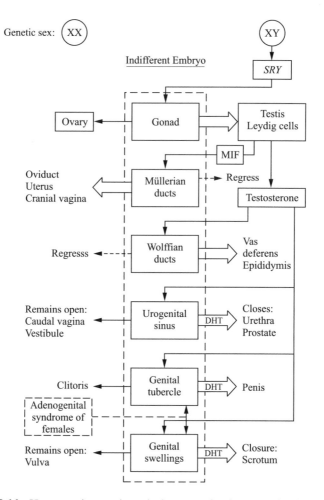

Figure 2.11. Hormone interactions during reproductive tract development. In males (XY), the müllerian duct regresses whereas the wolffian duct forms the epididymis, seminal vesicle, and prostate. However, a remnant of the müllerian duct contributes to a subregion of the prostate, the utricle (section 2.6.2). In females (XX), the converse occurs: the müllerian duct gives rise to the uterus and cranial portion of the vagina. In the adrenogenital syndrome, a genetic female fetus may be partially masculinized by adrenal androgens. DHT, dihydrotestosterone, the most active androgen which is enzymatically derived from testosterone; MIF, müllerian inhibitory factor; *SRY*, sex determining region of Y chromosome (formerly the *TDF* in humans and *Tdy* in mice). The mechanisms of testis determination are incompletely understood, but involve interactions of SRY with other genes, including MIF (also referred to as MIS) and the X-linked gene *Dax1* (Jiménez and Burgos, 1998; Hagg and Donahoe, 1998; Swain et al., 1998). This diagram does not show interactions with the developing nervous system, which can include demasculinizing and defeminizing mechanisms of fetal sex steroids. Revised from Meyers-Wallen (1993).

Gender differences in human brain preoptic neurons

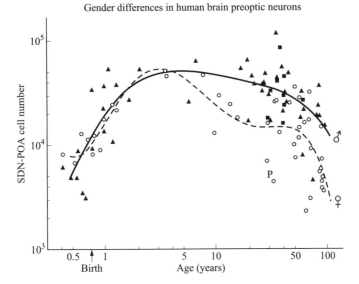

Figure 2.12. Quantitative variation in neuron number in the human male and female brain: the sexually dimorphic nucleus (SDN) of the preoptic region (POA) from fetus to old age plotted on a log-log scale. The SDN continues to add neurons until 4 years after birth, when differences begin to emerge in which the female nucleus becomes smaller, presumably through neuron death. Note the huge range of variations at all ages and the major loss at later ages. Open circles and dashed line, females; solid symbols and solid line, males; squares, homosexual orientation. P, Prader-Willi syndrome, with smaller numbers of neurons. From Swaab et al. (1992), with permission.

tion to making inferences across species about the importance of individual differences in sex steroid levels during development.

In view of these experimental manipulations of neuron number, it is of interest to consider the possible role of individual variations of fetal testosterone, which have ranges that overlap in males and females (Fig. 2.10A). In humans, fetal testosterone varies more than fivefold among individual males during the end of the first trimester when the testes are transiently activated (Reyes et al., 1974; Nagamani et al., 1979). Data of Ellinwood and Resko (1980) show that 46% of male fetuses had testosterone levels within the female range. Rodeck et al. (1985) also remarked on the overlapping range in fetal blood and amniotic fluid testosterone at midtrimester. Male rhesus monkeys (Resko et al., 1980) and rats (Weisz and Ward, 1980) show similar wide variations of blood testosterone after the onset of fetal testicular activity. Even later in gestation, when male fetal testosterone drops back into the lower range of females, both male and female human fetuses vary widely in both estradiol and testosterone (Fig. 2.10B). In contrast to testosterone, fetal plasma estradiol did not differ between male and female humans (Nayar et al., 1996; Reyes et al.,

1974) or rhesus monkeys (Resko et al., 1980). In each of these studies, the steroid assay *C-VAR*s were 10–20%, which is less than the greater than twofold differences among individual fetuses.

We suggest that these variations of sex steroid levels originate in individual differences in the numbers of hormone-producing cells in the fetal gonads, which are consequent to chance variations in PGC proliferation and survival in both males and females. The numbers of testosterone-producing Leydig cells per gonad would be expected to differ among individuals prenatally (we found no data in this regard). Serial sampling is needed to show if individual fetuses have sustained characteristic differences in sex steroid levels or if the variations are transient during chance activation of different numbers of ovarian follicles or Leydig cells, but this techniqiue is not feasible in humans because of risks to the fetus. Institutional review boards for animal experimentation in the United States are also reluctant to give permission for such studies on primates. However, noninvasive imaging techniques such as functional magnetic resonance imaging (fMRI) may become sensitive enough to resolve fetal gonad cell populations in humans and primate models.

The variations in fetal sex steroid levels in humans and primates suggest a role for the individual variations in development of the brain and other parts of the reproductive system (the prostate is discussed in section 2.6.2). The overlapping variations in hypothalamic neuron number between males and females (e.g., Figs. 1.8A, 2.12) could be consequent to the intrinsic variations in prenatal sex steroids being discussed. This concept is consistent with data from rodents and humans, discussed in Section 4.1 on fetal neighbor effects, in which modest variations in fetal sex steroid levels modify brain development with strong effects on adult functions. The individual variations in neuron numbers in the hypothalamus, corpus callosum, and hippocampus, and so on, could be pertinent to individual differences in adult behavior, as well as to outcomes of aging in sex-differentiated circuits, according to the threshold model for neuron loss (Fig. 1.9).

The individual variations in fetal sex steroid levels also pertain to the Geschwind-Behan-Galaburda model of cerebral lateralization (Geschwind and Galaburda, 1985a,b,c), which hypothesises that prenatal testosterone differentially modifies the growth and symmetry of cortical and subcortical structures, including the planum temporale. "Anomalous" cerebral dominance in left-handedness was associated with language and immune disorders. This model and the remarkable intellectual synthesis behind it have stimulated numerous studies, which, on balance, do not strongly support its main tenets (Berenbaum and Denburg, 1995; Hampson and Moffat, 1994; Forget and Cohen, 1994; Bryden et al., 1995; but see Clark et al., 1991). Moreover, recall the fluctuating asymmetry of the planum temporale in identical twins, which is independent of the co-twin concordance in handedness (Fig. 1.11). This dissociation of planum temporale asymmetry and handedness is also a minor challenge to the Geschwind-Behan-Galaburda model. Nonetheless, whatever the role of fetal testosterone in cerebral lateralization, associations are reported for brain asymmetry with individual differences in the peripheral immune response (Kang et al., 1991; Delrue et al., 1994; Neveau, 1993; section 1.5.3).

2.6.2. Prostate and benign prostatic hyperplasia

The prostate is another candidate for developmental effects of fetal variations that may modify the risk of benign prostatic hyperplasia (BPH). This condition occurs generally during aging in humans and in some other mammals (see Table 1.2). BPH arises in 50% of human males by age 60 and in > 90% by 80 years (Ahmed et al., 1997; vom Saal et al., 1994; Pike and Ursin, 1994). However, many human populations have not been characterized for BPH.

Benign prostate hyperplasia may begin soon after puberty, because some autopsies showed microscopic nodules in adolescents. The slowly growing and noninvasive nodules can cause urinary tract obstruction (clinical BPH) (Meikle et al., 1995, 1997) although lower urinary tract symptoms are not strongly correlated with the degree of BPH (Meickle et al., 1999). The development of BPH requires the presence of active gonads (it is not found in adult eunuchs), but the relationship to age changes in sex steroids is unclear.

The heritability of BPH in identical twins is low. As measured ultrasonically in 214 healthy twins, prostate volume had limited (22–30%) heritability (Meikle et al., 1997). Nonetheless, in this study and another questionnaire-based study (Partin et al., 1994), identical twins had a higher concordance for BPH than dizygous males. In the general population, familial risk factors are associated with about 15% of clinical BPH (Sanda et al., 1994).

We suggest that developmental variations in the prostate zone at risk for BPH be considered as distinct from the putative and unidentified environmental risk components. BPH mainly results from nodular growths in the peri-urethral glands and transition zone (*TZ*) (Fig. 2.13A,B,C, *TZ*), which comprise a small portion (< 5%) of the total gland in young men (McNeal, 1978, 1990; Timms, 1997; Verhoeven and Hoeben, 1994). vom Saal et al. (1994, p. 1225) pointed out the potential significance of nodules in the central zone, which may be subject to particular developmental variations as described below. The nodule-forming regions are distinct from and do not overlap with the locations of adenocarcinomas, which typically develop in the much larger peripheral zone (*PZ*) (Fig. 2.13A, *PZ*).

BPH is due to the focal formation of nodules of tissue in prostatic glands that complete their development at puberty. Most pathologists agree that BPH in humans is not a general or diffuse cellular growth process in the prostate. In McNeal's (1990, p. 479) description:

Human prostate zones and BPH

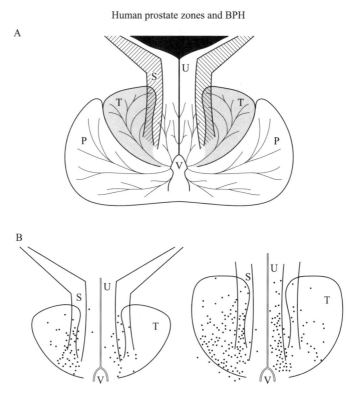

Figure 2.13. Human prostate anatomy, showing the subregions prone to development of benign prostate hyperplasia (BPH). From McNeal (1990), with permission. *A*, Locations of BPH in the periurethral region (U), transition zone (T), and the preprostatic sphincter (S), containing the most frequent locations of BPH nodules; V, verumontanum, an anatomic reference point. Together, the BPH prone regions account for < 5% of the volume of the normal gland. The much larger peripheral zone (P), where BPH nodules are rare, is the location of most cancers. *B*, Locations of micronodules early in PBH (*left*) and at later stages (*right*), superposed on A.

> a variable number of tiny orifices scattered along the urethral wall . . . give rise to abortively developed duct branches and acini . . . . Within this tiny volume, however, the locations of individual nodules [that arise during BPH] . . . appear to be randomly determined.

The de novo budding during early BPH arises through specialized growth processes, in which the microscopic nodules repeatedly rebranch through abnormal arborizations during BPH (McNeal, 1990; Timms, 1997). A difference from that of normal fetal development is that a BPH nodule may contain branches of more than one duct (McNeal, 1990, p. 480).

Fred vom Saal hypothesized that the size of the BPH-prone region and the risk of enlargement during aging are due to variations in exposure to hormones that influence growth of the BPH-prone subregions during development (vom Saal et al., 1994, p. 1227; vom Saal et al., 1997). First, we describe the extent of individual variations in healthy men without a clinical record of prostate disease. In two studies, prostate regional volumes were measured by ultrasound, which can accurately follow changes in prostate volume during treatment for cancer and BPH (see Table 2.1 notes). Co-twins from a Utah study population differed extensively in the size of the BPH-prone transitional zone, whether the twins are MZ or DZ (Fig. 2.14). Co-twin differences are apparent before 40 years and increase at later ages. It is likely that co-twin differences arise soon after puberty. Similarly, there are 10-fold differences in the transitional zones of young Danes from a general sample (Jakobsen et al., 1988). Prostate size is unrelated to body height or weight (Jakobsen et al., 1988; Meikle et al., 1995). Table 2.1 shows the C-VAR for the prostate volume in these studies and in a postmortem study based on dissection and direct weight. On the average, there is a 1.65-fold larger C-VAR for PBH-prone regions than for the total gland.

Table 2.1. Coefficients of variation (C-VAR) in the size of the prostate and the BPH-prone subregions

Method	Total prostate	BPH-prone subregion (alternate name)	C-VAR ratio, BPH-prone:total
Autopsy, sudden death (Leissner and Tissel, 1979b)	Wet wt.	Medial lobe, wet wt.	
16–19 yrs ($N = 8$)	17.6%	22.3%	1.27
20–44 yrs ($N = 48$)	20.7%	23.8%	1.15
45–59 yrs ($N = 17$)	16.3%	26.7%	1.64
Ultrasound (Jakobsen et al., 1988)	Volume	Peri-urethral gland, volume	
27–30 yrs ($N = 12$)	22.4%	32%	1.43
31-40 yrs ($N = 57$)	24.8%	47%	1.90
41–50 yrs ($N = 54$)	20.9%	44%	2.11
51–60 yrs ($N = 6$)	15%	23%	1.53
Ultrasound (Meikle et al., 1995)	Volume	Transition zone, volume	
27–77 yrs ($N = 428$)	5%	11%	2.2
Grand mean	17.8%	23.2%	1.65

BPH-prone subregions showed consistently greater variations (grand mean, 1.65) than those of the total prostate. Prostate disease was excluded. The medial lobe in Leissner and Tisell (1979b) includes the peri-urethral zone in Jakobsen et al. (1988) and the transition zone in Meikle et al. (1997a,b; Fig. 2.14). These variations represent tissue mass rather than prostatic fluid, because the water content varied much less (C-VAR, 1–2%). The transurethral ultrasound measurements (TRUS) were validated by comparisons with MRI and anatomical prostatectomy specimens (Rahmouni et al., 1992; Wolff et al., 1995). Reproducibility of prostate volumes in repeated tests is >97% (Rahmouni et al., 1992). Data from Leissner and Tisell (1979b) excluded specimens with macroscopic nodules.

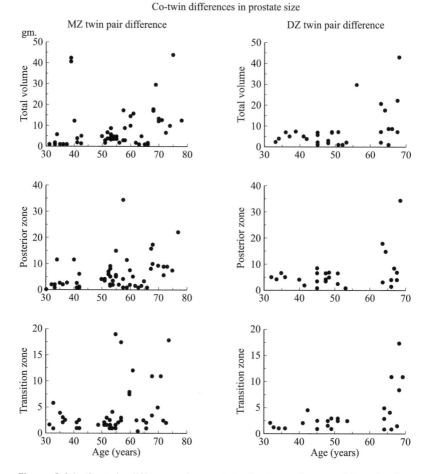

Figure 2.14. Co-twin differences in prostate size from ultrasound imaging in a sample of Causcasian men from Utah. The differences are proportionately larger for the small transition zone, which is the location of nodules of benign prostate hyperplasia (BPH). The differences between co-twins also increase with age to similar degrees in both MZ and DZ. Redrawn from Meikle et al. (1997a).

The extensive variations in adult prostate size suggest that they might be detected even earlier. The ultrasound data do not resolve variations before puberty, when the prostate is very small (4 g). Extensive fetal variations in early development are indicated in the sketches of Zondek and Zondek (1975), but histological studies are needed to define their extent.

Vom Saal hypothesized that subregional prostate size variations may arise through several mechanisms, each of which, as we note, is subject to chance. Size variations in the BPH-prone subregions may arise through chance variations in fetal estrogen levels during development. Fetal mice given estradiol implants that increased fetal blood estradiol by 50% also had 40% more prostate gland buds in late embryos (vom Saal et al., 1997). As adults, their prostates had threefold higher cellular levels of androgen receptors, which vom Saal hypothesized should increase the sensitivity of the adult prostate to androgens. By these experimental data, the fivefold variations of estradiol in human fetal blood of both males and females (Fig. 2.10) would be expected to cause quantitative differences in prostate buds and androgen receptor content. In turn, the greater androgen receptor content should support greater cell growth responses to testosterone at puberty and later when benign nodules begin to form. The fivefold variations in human fetal testosterone are 10-fold greater than the 50% range of values in mice given testosterone implants.

Variable müllerian duct tissue in the utricular plate could also contribute to prostate regional size variations (vom Saal et al., 1994, 1997). Müllerian ducts regress during male development through a hormone relay, in which fetal androgens from the testes stimulate müllerian inhibitory factor (MIF), which in turn induces müllerian duct regression through cell death (Fig. 2.11). Both testosterone and estradiol modify target cell responses to MIF, which gives one source of variation. Chance variations could also arise from the amount of MIF produced locally, as well as in the chance activation of cell death programs (see chapter 3), which in turn could lead to different numbers of surviving cells.

Returning to the data on twins (Fig. 2.14), the differences in prostate volume between individual twin pairs increase strikingly after 60. Longitudinal ultrasound measurements might reveal if the twin with the largest volume increase at later ages also had larger BPH-prone regions when young. Such data, if eventually obtained, would help evaluate the extent of postnatal environmental factors in BPH. Remarkably, no environmental or lifestyle risk factors for BPH have been established across human populations (Ahmed et al., 1997; Meickle et al., 1997a,b).

Prostatic adenocarcinoma, which is much more dangerous than BPH, arises in a different subregion, the larger peripheral zone that comprises 90% of the total volume. Prostate cancer is the most commonly diagnosed cancer in men, with an increased incidence about 15 years after BPH (Pike and Ursin, 1994; Meickle et al., 1995). Metaplastic cells are found in the prostate of 40% of men by 50 years and 70% by 80 years (Pienta and Esper, 1993; Ahmed et al., 1997). The lifetime risk of developing invasive prostatic cancer is about 15%, with risks increasing exponentially after 40 years. Overall, < 10% of prostatic cancer is considered heritable (Page et al., 1997), which implies a role of chance developmental variations, for example, through chance variations in cell numbers or cell–cell interactions. In view of evidence for altered epithelial–stromal interactions in abnormal prostate growth (Lee, 1997; Tenniswood, 1986), we suggest that chance variations in the numbers of stromal and epithelial cells acquired during development may also determine which branches acquire abnormal growth regulation.

2.7. Prezygotic phenomena

A little-recognized topic in life history is chance influences on the oocyte from which an individual arises before fertilization, hence the term "prezygotic." We considered earlier the maternal age effects on fetal aneuploidy (section 2.4), which for Down syndrome are mainly attributed to nondisjunction before fertilization. Another, also specialized example is the Lansing effect observed in certain rotifers and crustacea. In rotifers (*Philodina citrina* and *Euchalis dilatata*) and a water flea (*Moina macrocopa*), the offspring of eggs shed by older females had shorter life spans, the "Lansing effect" (Lansing, 1947; Murphy and Davidoff, 1972; reviewed in Finch, 1990, pp. 484–486). With increasing maternal age, the rotifer eggs became more variable in size and more colored, which suggests a cytoplasmic determinant, possibly originating in the ovarian nurse cells (Finch, 1990, p. 486). Another prezygotic phenomenon may be found in the persistent effects of heat shock on fertility in *Drosophila*, which can extend for two generations (Bucheton, 1978). These examples merit reexamination with the wonderful molecular tools now at hand for analyzing the oocyte and early embryo, as described in chapter 3.

In avian eggs, sex steroid levels determine social and reproductive behaviors. In this case, the differences in fetal steroids are not due to the embryo but prexist in the egg before fertilization. In each clutch of canary eggs (*Serinus canaria*), the amount of maternal testosterone in the egg yolk increases with the order in which the eggs are laid, over a fourfold range of testosterone concentrations (Fig. 2.15A; Schwabl, 1993, 1996). The yolk testosterone correlated with maternal blood testosterone (Schwabl, 1996). In turn, the social rank of the juvenile canary (Fig. 2.15, B) and the aggressiveness of the adult (Schwabl, 1996) are correlated with the amount of egg testosterone. Although these mechanisms have not been defined, there are good precedents in birds for effects of sex steroids on neuron numbers during development, which have defined effects on adult behaviors. The role of chance in these processes may be evolutionarily adaptive. Because canary eggs hatch asynchronously about one day apart, Schwabl (1993) suggested that the higher testosterone content of the later eggs may support faster maturation of the motor system, which might compensate for the later hatching. Moreover, maternal testosterone is influenced by population density and other environmental factors (Wingfield et al., 1990), which thus can be transmitted as a transgenerational signal to optimize individual development to local conditions (Schwabl, 1996). Overall, we note that this mechanism yields multiple behavioral phenotypes within the same clutch of eggs from a single mated pair, which may increase the chance of success of offspring

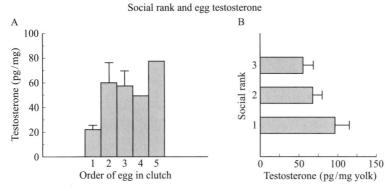

Figure 2.15. Social rank in juvenile canaries is associated with higher levels of testosterone in the egg at the time it was laid. *A*, The first eggs have lower testosterone content. *B*, The social rank of adolescent birds parallels the amount of testosterone in the yolk. Redrawn from Schwabl (1993).

with different birth order to contend for varied reproductive opportunities according to fluctuations in social structure.

We return to mammals, by noting that the prezygote egg is subject to transgenerational influences that extend back to the *grandmaternal* environment (Finch, 1996; Finch and Loehlin, 1998; see Table 1.1). As described above, oogenesis in mammals is complete at birth. that is, the cell that we grew from existed in our mother's ovaries at her birth and thus could be subject to chance environmental influences that determined the environment of the grandmaternal uterus. In humans, elephants, and certain whale species, an oocyte could be more than 50 years old. The possibility of prezygotic variations was also considered in efforts to explain morphological variations within litters of mice that did not diminish during extensive inbreeding and differences between artificial MZ twins in mice (Gärtner and Baunack, 1981; Baunack et al., 1986, 1988; Gärtner, 1990; Note, p. 75).

As a first approach to this question in humans, cognitive and behavioral traits were examined in MZ twins in relation to maternal age (Finch and Loehlin, 1998). The rationale was that oocytes from older mothers would have been exposed to greater risk of damage from free radicals, which might have subtle effects on brain development. However, no effect of maternal age was observed. Other traits from the large twin databases are being examined.

These examples suggest that prezygotic influences will vary widely among species, according to the physiology of oogenesis. In view of the highly developed techniques for ovum transplantation, direct tests of prezygotic damage could be designed by manipulating the grandmaternal environment. According to its evolutionary history, a species may have evolved specialized mechanisms to minimize damage during the often very prolonged sojourn of the oocyte in the ovary before fertilization. The studies of avian sex hormone signals transmitted in the egg may differ among species in the way different kinds of environmental fluctuations modify outcomes of development. The greater aggressiveness of adult birds from eggs with high testosterone, for example, would be expected to influence mortality risks and life expectancy.

2.8. Conclusions

We considered how chance variations during development of the reproductive system have influences on outcomes of aging. The case is strongest for females. Experimental reduction of the ovarian mass shows that the number of ovarian oocytes present in a young mouse is the major determinant of when fertility is lost during aging. We conclude that the severalfold variations among humans in the stock of ovarian oocytes present at birth or in a young adult are the major determinant of the age of menopause. Moreover, the maternal age risk of Down syndrome and other fetal aneuploidies can also be attributed to the size of the oocyte pool, because removal of one ovary in mice increases the incidence of fetal aneuoploidy. Whenever menopause occurs (naturally or as the result of surgery), the resulting loss of estrogens sharply increases the risk of cardiovascular disease, osteoporotic fractures, and possibly also Alzheimer disease. We extended these questions to a common outcome of aging in men, benign prostate hyperplasia (BPH). Although the information in this area is much less developed, several lines of evidence suggest that there are extensive chance developmental variations in the amount of tissue in the subregion of the prostate in which BPH arises. Again, the heritability of prostate regional volume is relatively low, consistent with the hypothesis. The variations in oocyte number and, possibly, in the size of prostate subregions appear to be present from early development. Further studies are needed of the control of cell number variations in the reproductive system, including interactions between fetal steroid fluctuations and the brain.

3

Chance, Cell Fate, and Clonality

A fundamental aspect of multicellular life is the interplay between cell lineage and clonality. In vertebrates, an individual begins as a single cell—the zygote—and then, through a series of cell divisions and differentiation events, effectively becomes a large, variegated cell clone that itself comprises a mosaic of innumerable subclones (Davidson, 1990; Wolpert et al., 1998). Not only do cells undergo division and differentiation during development, but also in adult cells may change their phenotypes by interacting with neighboring cells, may migrate, or may die. As the overall space-time cell lineage of the organism unfolds, cells are subject to a variety of stochastic processes that may affect their fate. We now examine how variations in cell fates arise during development and adulthood and how these variations may contribute to the unfolding of aging phenotypes.

3.1. Stochastic aspects of cell fate

3.1.1. Asymmetric cell division

The number and type of cells in a lineage of differentiating cells can be subject to chance fluctuations because of asymmetric cell division, which differs with cell lineage and species and can be very deterministic. During cell replication in insect, sea urchin, and vertebrate embryos and so on, daughter cells may be subject to a characteristic probability for their future fates that lead to different (hence asymmetric) cell specifications. Cell fate differences among daughter cells may not be obvious by the initial microscopic appearance, although in some cases the daughter cells are structurally different. In certain cell lineages, differences among daughter cells can arise from different planes of mitotic cleavage, which may partition subcellular materials that were concentrated or localized to one end of the mother cell, or leave just one daughter cell attached to a specialized cell or extracellular environment (Fig. 3.1).

Asymmetric cell fate assignment occurs in two major categories of body cells and elements: those that can be regenerated after maturation and those that cannot. As described in chapter 1, many body elements are fixed in number by the time of maturation and are irreplaceable, such as nondividing cells (oocytes, neurons) and the size and number of certain multicellular organizations (brain gyri, bristles). A second category of cells and elements is those that are replaceable. During cell turnover, the number of cells replaced is also subject to chance fluctuations through asymmetric cell divisions. Sometimes in the embryo and adult, one or both daughter cells retain the capacity for extensive or unlimited proliferation, as can be demonstrated in special cases that are referred to as stem cells.

In many examples, asymmetric division is governed by the geometry of cell-to-cell contacts, that depend on the local architecture of the tissue. The *Drosophila* ovary illustrates how the local geometry of cell position leads to asymmetric divisions (Fig. 3.2). The apical tip of each ovariole contains two to three germline stem cells lying just below apical filament cells that regulate the proliferation of germline stem cells (Lin and Spradling, 1993). Eggs mature on an assembly line, moving away from the apical tip toward the oviduct. The germline cells are the sole source of the eggs and are exclusive descendants of the special lineage of primordial germ cells (PGCs) that is differentiated in the early embryo (chapter 2.3). One cell retains extensive (possibly unlimited) capacity for proliferation as a stem cell, whereas the other daughter cell is committed to differentiate as an oocyte. PGC stem cells have a specialized subcellular organelle, the spectrosome, at the apex of the cell next to the terminal filament cells (Lin and Spradling, 1997; Lin and Schagat, 1997). This asymmetric location allows the spectrosome to anchor the mitotic spindle. Consequently, the plane of the cell at mitotic cleavage is set so that one pole of the mitotic spindle is always associated with the spectrosome and the terminal filament cells. As stem cells divide, one daughter cell remains in contact with the apical filament cells through the anchor of the mitotic spindle to the spectrosome. The other daughter, a future oocyte, does not contact the apical filaments. In this case, the daughter cell fate in asymmetric division is determined by a unique cytoplasmic organelle, which in turn dictates the architecture of cell-to-cell contacts.

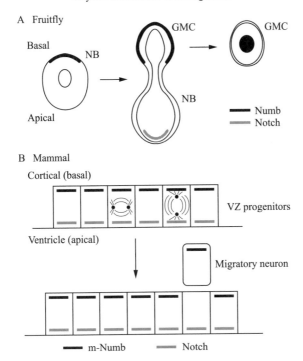

Asymmetric cell fate in neurogenesis

A Fruitfly

Basal

NB

GMC

GMC

Apical

NB

━━ Numb
━━ Notch

B Mammal

Cortical (basal)

VZ progenitors

Ventricle (apical)

Migratory neuron

━━ m-Numb   ━━ Notch

Figure 3.1. In the developing CNS of fruitflies and mammals, neuroblasts proliferate in a sheet of epithelial cells, in which there is an apical–basal polarity. Although there are important species differences, a shared feature in flies and mammals is that the daughter cells from asymmetric divisions have different amounts of a key protein Notch which is antagonized by Numb (see below) in regulation of the Delta-Notch signaling pathway (Buescher et al., 1998; Skeath and Doe, 1998; Wakamatsu et al., 1999). *A*, Fruitfly. The cleavage plane during neuronogenesis appears to be less precisely maintained than in the fruitfly ovary, which may allow statistical fluctuations in the partitioning of Notch among daughter cells, *B*, VZ, ventricular zone. From Zhong et al. (1996, Fig. 4d), with permission. *B*, Mammal. In the developing mammalian cerebral cortex, Notch is found at the base of the cells, which determines in large part the outcome of cell division, because neuroblasts largely divide at mitosis in two geometric planes. If the cell division is *vertical*, this cleavage plane more-or-less evenly partitions the Notch depot into the daughter cells, both of which remain in the epithelial layer and continue to proliferate (symmetric cell division). However, if the division is *horizontal*, the basal daughter gets all or most of the Notch depot and then migrates away from the neuroblast layer and begins differentiation (asymmetric cell division). Numb is asymmetrically partitioned at division into the ganglion mother cell (GMC) and is absent in the neuroblast (NB). From Lin and Schagat (1997), with permission.

Fruitfly ovary

A Ovary

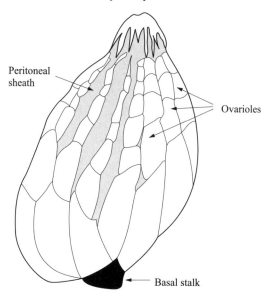

Peritoneal
sheath

Ovarioles

Basal stalk

B Ovariole

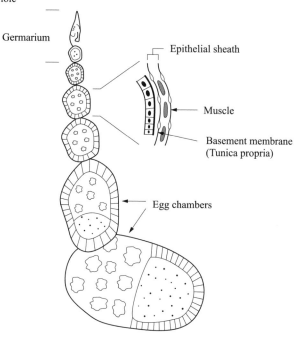

Germarium

Epithelial sheath

Muscle

Basement membrane
(Tunica propria)

Egg chambers

Neurons are also formed through asymmetric cell division. As in many other tissues of embryos and adult organisms of flies and vertebrates, the fate of neuronal progenitor cells (neuroblasts) is determined by the local tissue architecture through extrinsic spatial cues. In the developing CNS of flies and mammals, neuroblasts proliferate in a sheet of epithelial cells, which has an apical-basal polarity (Lin and Schagat, 1997; Chenn and McConnell, 1995; Zhong et al., 1996). In flies and mammals, asymmetric division partitions different amounts of key signaling system effectors, the proteins Notch and Numb. Experiments with *Drosophila* embryos show that neuronal fate is determined by the antagonism of the Delta-Notch signaling pathway by Numb (Fig. 3.1, legend). In the developing mammalian cerebral cortex, Notch is found at the base of the cells. This location determines in large part the outcome of cell division to cell fate, because neuroblasts generally divide at mitosis in either of two planes or orientations. If the cell division is vertical, this cleavage plane more-or-less evenly partitions the Notch protein depot into the daughter cells, which both remain in the epithelial layer and continue to proliferate (symmetric cell division). However, if division is horizontal, the basal daughter gets all or most of the Notch depot and then migrates away from the neuroblast layer and begins to differentiate (asymmetric cell division). The cleavage plane in the developing brain may be less precisely maintained than in the fruitfly ovary, which could lead to variable partitioning of Notch among daughter cells. The impact of such deviations on daughter cell fate is not yet described. Notch signaling systems have general importance to many species. One mechanism is hypothesized to be chance signaling fluctuations that, as cell colony size increases, yield varying degrees of "lateral inhibition" between neighboring cells (Heitzler and Simpson, 1991; Kopan and Turner, 1996).

Many cell types display asymmetric division in the absence of apparent extrinsic signals. In culture, clones from single cells produce diverse cell types. For example, clones of single neurons from vertebrate embryos give rise to mixtures of cell types in the resulting colony, including neurons and the two main types of neuroglia, astrocytes and oligodendroglia (e.g., Semple and Anderson, 1992; Kilpatrick and Bartlett, 1993).

Figure 3.2. The ovary of the adult fruitfly illustrates how different planes of mitotic cleavage may partition subcellular materials that were concentrated or localized to one end of the mother cell. *A*, The ovary showing the ovariole subunits. *B*, Each of the ovarioles within an ovary contains a string of six or seven eggs at different stages of maturation. The tip of each ovariole contains stem cells that continuously produce new egg chambers. Each egg chamber contains one oocyte and 15 nurse cells that pump yolk and other materials into the maturing egg, all surrounded by about 1000 follicle cells. The maturing eggs move from the apical tip to the oviduct, from which they are deposited. From Spradling (1993), with permission.

Even adult brains have subpopulations of multipotent progenitor cells that can be stimulated to proliferate and to differentiate into a variety of brain cell types (Gage et al., 1998). The extent of proliferation and the cell types produced in these cloning studies are influenced by growth factors (e.g., epidermal-, fibroblast-, and neuron-growth factors) and extracellular matrix molecules (e.g., fibronectin). Regions of the vertebrate CNS vary extensively in the proliferative potential. For example, although motor neurons of different parts of the chick embryo spinal cord are superficially indistinguishable, there are twofold regional differences in neuron number within the developing spinal cord of individuals even before the onset of neuron death (Oppenheim et al., 1989).

These studies and many others show wide statistical variations in the numbers of each type of cell produced. Two general working hypotheses guide experimental design and interpretation of data: (i) the outcomes of asymmetric division are subject to chance (stochastic) processes, and (ii) the range of potential phenotypes is progressively restricted, as binary choice points are passed that restrict the potential numbers of genes that could be activated. Although the term "stochastic" is freely used in this literature, few studies have acquired sufficient data to test for genuine stochasticity.

3.1.2. Cell migration

Chance further operates on numbers that may vary from indeterminacy during cell migration. In early development of most animals, there is a phase of extensive cell proliferation, followed by massive and complex movements of cells, either as sheets of cells, or as individual cells from one part of the embryo to another. Even adult cells may migrate considerable distances, as is generally true for blood cells formed in the bone marrow, which in some cases must pass through the thymus or other specialized tissues to complete their differentiation. In the case of neurons, even if the cell body and cell nucleus do not move, axons and dendrites can extend or retract over considerable distances to vary cell contacts. These migrations of cell bodies or cell processes are dependent on molecular signals, and thus are subject to chance variations in the surrounding molecules and cells.

Nervous system development in vertebrates depends on massive cell migration, first to form the neural crest that defines the longitudinal axis of the future nervous system. Then, neural crest cells from asymmetric divisions migrate out into future locations of subregions and differentiate as outlined in the preceding section. During these complex migrations, some cells fail to reach their correct location (ectopic cells). A well-studied example is in the isthmo-optic nucleus (ION) visual system of the chick embryo brain, where about 2–3% of ION neurons are located in the zone around the main neuron cluster (Fig. 3.3). Most of these ectopic cells die before brain maturation is complete, as described below. Although cell death, as discussed below, accounts for much of the decrease in neuron number during development, there may also be migration of postmitotic neurons. Exacting analyses with several biochemical markers would be required to resolve these processes.

3.1.3. Cell death

Many cells may die after they differentiate through programmed cell death (apoptosis), in which extrinsic cues again depend on chance environmental encounters. Neurons and other cells die on a massive scale during development; in most tissues, the range is 40–90% (Saunders, 1966; Jacobson, 1991; Oppenheim, 1991). Two main forms of cell death are apoptosis and necrosis. In current thinking, apoptosis is programmed cell death, whereas necrosis more typically occurs during tissue injury and is associated with inflammation. The two modes of cell death were suggested from cell morphology (Wyllie et al., 1980; Searle et al., 1982). Apoptosis is widely described during development and differentiation. This mode is characterized by cell shrinkage (pyknosis), with condensation and fragmentation of the nucleus and breaking of DNA strands, which is histochemically assayed by end-labeling of DNA, as an assay for apoptosis. Cell fragments may be phagocytosed by neighboring cells in the embryo, rather than invading professional phagocytes such as macrophages. Apoptotic cell death usually occurs in single cells, whose neighbors remain viable. Necrosis is associated with cell swelling and more rapid damage to mitochondria. Necrotic cell death during mechanical injury or infection is associated with the immigration of macrophages and other inflammatory cells. Clusters of cells may die through necrosis. The morphological and biochemical characteristics of cell death vary widely among cell types and tissue age and state, which has led to the proposals for further subtypes of cell death (Clarke, 1990; Evan and Littlewood, 1998; Thornberry, 1998). Ultimately, the sets of genes that are programmed during differentiation to be expressed in a particular cell type must determine the details of change during cell death. Cell death can be modified by interactions with neighboring cells, and is thus not a hard-wired program. It seems likely that all modes of cell death will prove to share some mechanisms and pathways.

A well-quantified example of neuron death occurs in the avian ION, which sends axons from the brain to the retina of the opposite eye (Fig. 3.3). About 60% of the 22,000 ION neurons that are initially formed die during the week before hatching (Clarke and Cowan, 1975, 1976; Clarke et al., 1976). Retrograde tracing studies show that nearly all (> 95%) of initially formed ION neurons synapse with the retina; thus, afferent connections alone do not ensure neuron survival (Clarke and Cowan, 1976). The final number of surviving neurons in this inbred avian strain varies by individual with a range of 10–15%, which Clarke et al. (1976) describe as "fairly constant." Two small subpopulations of aberrant neurons

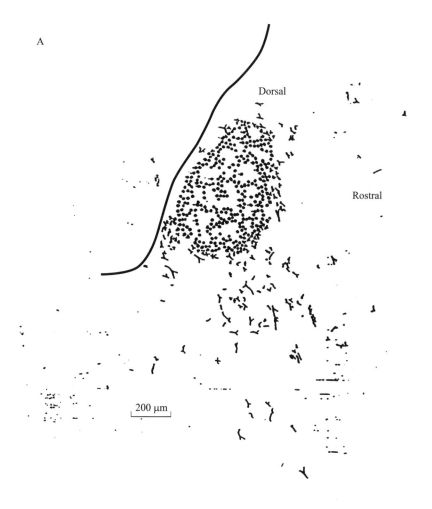

A

Dorsal

Rostral

200 μm

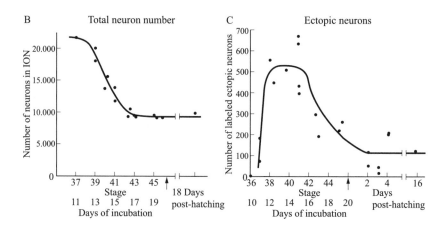

B   Total neuron number

Number of neurons in ION

20.000

15.000

10.000

5.000

0

Stage 37 39 41 43 45

Days of incubation 11 13 15 17 19

18 Days
post-hatching

C   Ectopic neurons

Number of labeled ectopic neurons

700

600

500

400

300

200

100

0

Stage 36 38 40 42 44 2 4 16

Days of incubation 10 12 14 16 18 20

Days
post-hatching

show even greater loss (Clarke et al., 1976). Within the developing ION, about 0.5% of neurons send axons to the wrong (ipsilateral) eye, and none of these survive. The ectopic ION neurons (about 3% of the total) also undergo extensive, 80–90% loss. Nonetheless, some of these neurons survive.

Neuron death can be manipulated in many areas of the nervous system. If one eye is ablated in the embryo, then the loss of the main cluster of ION neurons is much greater than in unoperated controls (Clarke et al., 1976). Many other neural systems show that loss can be increased by removal of afferent connections or can be rescued (Jacobson, 1991; Oppenheim, 1991). For example, in the embryonic chick spinal cord, about 40% of motor neurons die. Of these, 25% can be rescued by grafting embryos with extra limbs (Hollyday and Hamburger, 1976). Motor neuron death is completely suppressed by treatment with bungarotoxin, a blocker of the neuromuscular junction (Oppenheim, 1991). The mechanisms that regulate the statistics of cell death are not known. In these and many other systems showing neuron death, a range of neurotrophins can rescue neuron cell death, for example, classic studies of Levi-Montalcini and Hamburger that identified the first of these proteins, nerve growth factor (NGF). Whereas NGF rescues some neurons, it induces death in oligodendrocytes (Casaccia-Bonnefil et al., 1996). Moreover, the geometry of cell contact with the surface substrate can determine whether a cell dies or proliferates, as elegantly shown in capillary endothelial cells that were grown on substrates with micropatches of fibronectin of different spacing (Chen et al., 1997). Thus, random-walk heterogeneity in the diffusion of extracellular trophic factors or matrix molecules has the potential to produce a virtually infinite array of different microenvironmental signals for cell death versus survival, as well as many subtypes of cell death.

Cell death continues in tissues of adults. In tissues that maintain dividing cells, cell death may occur by the shedding or sloughing off of cells, as in skin and mucosa; through apoptosis during cell differentiation, as in bone marrow; and through clearance of systemic cells with longer, but finite life spans, as in erythrocytes and platelets. In tissues with irreplace-

Figure 3.3. Cell death in the chick brain isthmo-optic nucleus (ION). *A*, The main ION neurons and ectopic neurons. The ION neurons project to the contralateral eye, which can be retrogradely labeled by intraocular injection of horseradish peroxidase (HRP). In 15-day chick embryos, 95% of ION neurons, including the ectopic cells, are labeled by HRP, including those that lie as far as 800 μm outside of the main nucleus. Most ectopic neurons will die within a week. *B*, Of the 22,000 ION neurons formed, about 40% die within the week before hatching. *C*, Ectopic ION neurons show greater proportional loss, but with more extensive variations among individual brains in the numbers that survive. From Clarke and Cowan (1976), with permission.

able cells, cell death occurs with high tissue specificity. The ovary is an extreme example, with the ongoing atresia of oocytes and follicles at a statistically regular rate (Fig. 2.3). However, in the brain, the extent of normal neuron death, that is, independent of specific diseases, is controversial (see section 1.4.2). The mechanisms of cell death in these widely different situations are likely to be very diverse (Thornberry, 1998; Evan and Littlewood, 1998).

Observations of cell death in inbred chicks and mice show substantial variability within and between individuals. Although many studies provide standard errors for numbers of surviving neurons, we do not know of an in vivo study that partitioned this variance into the initial variance of cell number (prior to cell death) versus variance due to cell death.

3.2. Stochastic aspects of cell clones

In chapter 2 (Fig. 2.7), we discussed extensive differences in proliferation among individual primordial germ cells (PGCs), suggestive of a stochastic process. We now look at other examples of clones in the brain, intestine, and bone marrow and in cultured fibroblasts in which the stochastic features have been studied in more depth.

3.2.1. Dispersion of neurons during cortical development

The cerebral cortex, which includes the neocortex, can be considered as equivalent to a folded sheet, with subareas devoted to vision, sensation, hearing, and other functions. Each area shows regional specialization of a basic cellular structure, comprising six neuronal layers. Furthermore, each subarea can be divided physiologically into smaller units, or columns, in which the six layers share anatomical connections and physiological properties. Thus, division of the cortex into "cytoarchitectonic" areas appears to be a basic organization of cortical function. How this organization is first established is not clear.

An early hypothesis was that the borders between cortical areas are defined by cell lineage during brain development. Thus, progenitor cells would give rise to daughter cells that were already committed to form part of a particular cytoarchitectonic area in the mature brain, and postmitotic neurons destined for different areas would be proscribed from mixing. Cell lineage studies using intracellular dye injections have shown that in the chick embryo, certain regional boundaries in hindbrain are indeed defined in this way (Fraser et al., 1990). The hindbrain is formed from segmental bulges (rhombomeres), which confine clones of cells, even if polyclonal in origin (Fraser et al., 1990), and which comprise distinct domains of differential gene expression (Nittenberg et al., 1997). A model of how cell lineage boundaries in the cortex might be translated into cytoarchitectonic area boundaries was defined by Rakic (1972, 1991), who suggested that a protomap of cortical areas was present in the germinal layer of the ventricular neuroepithelium, and that the neurons which will belong to different areas derive from a mosaic of distinct precursor populations. However, subsequent studies using markers for cell migration clearly show that neuronal clones become dispersed across cortical boundaries, with some degree of mixing (Crandall and Herrup, 1990; Walsh and Cepko, 1992; Grove et al., 1992).

Grove et al. (1992), including one of us (T.B.L.K.), analyzed clonal dispersion of neurons in the hippocampus, a subcortical structure in which the transitions between adjacent cytoarchitectonic areas are sharply defined (Fig. 1.5). A suspension of a retroviral vector, carrying the bacterial *lacZ* gene as a histochemical marker, was injected into the cerebral vesicles of rat embryos, which infected dividing ventricular zone cells and their progeny. Injections were made on embryonic day 16 (E16), and the brains of injected animals were analyzed during the third postnatal week when hippocampal area boundaries are mature. The viral titer was adjusted to yield an average of two neuronal clusters per hippocampus; at this level the likelihood of a chance superimposition of two clones in any one hippocampus was calculated as <2%. Out of a total of 126 neuronal clones (49 hippocampi), just under half were single cells, while the rest ranged from two to six cells. The multicell clones showed considerable uniformity in their spatial distribution, the neighbors in a clone typically separated by 100–200 μm (Fig. 3.4).

The dispersion of neuron clones was determined in the clear boundaries of the hippocampal formation that separate the subiculum, prosubiculum, and the four zones of the pyramidal neuron layer of the hippocampus, as named after Cajal (CA_1-CA_4) (Fig. 1.5). About 20% of the clones included cells within a few hundred microns of cytoarchitectonic area boundaries. Of these, <10% crossed boundaries (Fig. 3.5). A probability model for random dispersion of clones with and without a constraint on border crossing showed that it was highly unlikely ($P < 10^{-6}$) that the apparent border-crossing clones could have arisen by chance juxtaposition of separate clones on either side of a boundary. Border-crossing clones were observed as frequently as expected if clones spread freely over the hippocampus with no constraint imposed by cytoarchitectonic areas; the clones contained different types of pyramidal neurons characteristic of their locations. These data clearly supported the hypothesis that neuronal precursor cells generate clones that contribute to more than one cytoarchitectonic area in the hippocampal formation of the rat.

Retroviral marking techniques showed that neuronal clones are even more widely distributed across areas of the cerebral cortex (Walsh and Cepko, 1992). A library of 100 retroviruses was generated by inserting DNA fragments, distinguishable by PCR, into a viral vector containing the bacterial *lacZ* gene, and mixed suspensions containing these viral markers were injected into the lateral ventricles of late rat embryos. The number and distribution of labeled cells were recorded in three-

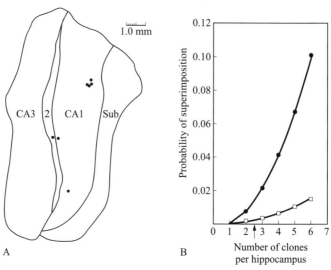

Figure 3.4. Evidence that the chance superimposition of two or more retrovirally labeled clones is highly improbably with a low density of infective events. *A*, Density of neuronal clones in a typical hippocampus. (To show the entire hippocampus in a single drawing, the principal cell layer has been "unfolded" and represented in a flattened reconstruction; the cytoarchitectonic areas shown here are the subiculum and CA1, CA2, and CA3 of Ammon's horn.) Three widely spaced clones are seen. One consists of a single neuron (dot); the other two are small clusters of two and four neurons. *B*, The probability of superposition in hippocampi containing varying numbers of clones, calculated according to probability models in which clones are assumed to be equal in area to the average seen in a sample of 49 hippocampi (upper curve) or to the largest clone ever seen (lower curve). From Grove et al. (1992), with permission. For anatomical circuits, see Figure 1.5.

dimensional reconstructions in postnatal brains. Clonal analysis was performed by amplifying the genetic markers with PCR from each single, histochemically labeled cell. For each clone, defined as a set of labeled cells showing the same marker upon PCR amplication, cell type was identified as neurons or glia, with most clones containing one or the other cell type, but not both. The neuronal clones formed two spatial patterns: clusters and nonclustered, or scattered, neurons. Clustered neuronal clones, defined as two or more cells (maximum cluster size of 6) with the same genetic tag located within 1.5 mm of one another, generally comprised groups of cells that were each others' nearest histochemically labeled neighbors. Neurons in these clusters were confined to one or two cortical laminae and often had similar morphological specialization. Clustered neuronal clones were not, however, limited to single functional subdivisions of the cortex, again showing that neuron precursors generate clones that contribute to multiple brain areas.

Clones spanning cytoarchitectonic boundaries

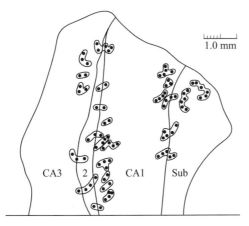

1.0 mm

CA3   2   CA1   Sub

Figure 3.5. A composite drawing of 25 retrovirally labeled clones in a sample of 49 hippocampi that lie within a few hundred microns of a boundary between cytoarchitectonic areas. Statistical analysis strongly rejected the hypothesis that the apparent-border-crossing clones could have arisen from chance juxtaposition of two clones on either side of the borders ($p < 10^{-6}$). From Grove et al. (1992), with permission.

Cytochrome oxidase histochemistry of the somatosensory cortex (Wong-Riley and Welt, 1980) reveals functional subdivisions of the somatosensory cortex, which form a systematic map of the contralateral body surface. Subdivisions, called "cortical barrels," reflect the pattern of whiskers on the face of the rat, with each barrel receiving information primarily from one whisker (Welker, 1976). Similar subdivisions in the sensory representations of the limbs correspond to single pads on the digits (Dawson and Killackey, 1987). When Walsh and Cepko (1992) analyzed the distribution of clonally related cells relative to the barrels, they found that none of the four clustered neuronal clones that occurred within the barrel field was confined to a single barrel, and all clusters crossed one or more barrel boundaries.

Moreover, many cells containing the same tag were scattered over large distances, sometimes approaching the largest dimension of the entire neocortex (Walsh and Cepko, 1992). These nonclustered cells were neurons, or a mixture of neurons and some unidentified cells, scattered singly or as several clusters. It is possible that some nonclustered cells carrying the same genetic marker arose from multiple infections of precursor cells. However, Monte Carlo simulations of the likelihood of such events indicate that some and probably most of the nonclustered patterns represent dispersion during migration of clonal descendants across much of the neocortex (Walsh and Cepko, 1993; Kirkwood et al., 1992; Walsh et al., 1992).

In conclusion, the studies of clonal dispersion of cortical neurons in the rat indicate that the patterns of cell generation during the clonal expansion and migration from germinal layer of the neuroepithelium (from about embryonic day 16) do not specify the later functional organization of the cortex into cytoarchitectonic areas. Neuronogenesis may produce a cortex that is initially functionally equivalent, which is later sculpted into distinct functional areas by cell–cell interactions. This developmental strategy, Walsh and Cepko (1993) suggested, must be an evolutionary advantage by allowing fast, partially independent evolution of the cortex within the body plan. This argument, although teleological, is consistent with the co-twin variations in cortical gyri (see Fig. 1.7). Moreover, we surmise that this model of neurogenesis also confers the advantage of reduced vulnerability to accidents in the fate of individual clones, such as somatic mutations, because of its polyclonal nature.

The dispersion of neuronal clones should also be considered in light of other factors that may affect the *numbers* of neurons formed in individual brain regions, particularly in relation to the neuron number fluctuations (section 1.4.2). In the early embryo neural tube stage, cells are proliferating that give rise to neurons and various glia (astrocytes and oligodendroglia). The frequency of horizontal divisions increases to provide the vast numbers of neurons and glia needed (Chenn and McConnell, 1996). Many more cells are produced than are eventually incorporated into adult brain structures, with excess cells presumed to give a margin of safety. At later stages in the cerebral cortex, neuron production almost completely ceases. We do not know the contributions of cell proliferation versus cell death to the variations of neuron number in adults, which differs between developmental stages for each brain structure and subregions; for example, the fraction of apoptotic cells with fragmented DNA (TUNEL positive) differs up to 5-fold between adjacent layers of the cerebral cortex of the neonate (postnatal day 1; Spreafico et al., 1995).

The fates of blast cells that migrate away from the proliferating zone are highly dependent on local influences during subsequent asymmetric divisions (Leber et al., 1990; Götz et al., 1995; Baker et al., 1997). By clonal expansion, neural tube cells produce a range of cell types including neurons (both motor and interneurons), astrocytes, oligodendrocytes, and ependymal cells. This was shown by labeling of individual cells that migrate away and become incorporated into specific pathways (Leber et al., 1990). Asymmetric divisions that produce multiple cell types continue throughout neurogenesis. Clones of rat cerebral cortex cells contained mixtures of cell types, including neurons with either of two main neurotransmitter types (GABA or glutamate, ratio of 1:4; Götz et al., 1995). Asymmetric cell divisions leading to different cell fates can occur up to the last division cycle. At later stages, the incidence of neuron death was similar in all clones migrating out from the neural tube (Leber et al., 1990). Whether the clones contained neurons exclusively or contained mixtures of cell types, 40% of neurons were lost, which approx-

Table 3.1. Intrabrain coefficients of variation (C-VAR) in cell numbers in postnatal rat somatosensory cortex

Postnatal day	Layers II–IV (cortical plate)	Layer V	White matter
1	13%	26%	13%
5	15%	26%	18%
14	38%	16%	65%

Mean total cell number: 3890

Calculated from Spreafico et al. (1995) by C.E.F.
Cell numbers were counted in three nonadjacent sections from one rat. The fraction of apoptotic (TUNEL-positive) cells was reported without statistics. The cell type was not specified and could vary with stage of development; white matter cell bodies are mainly glia.

imates the average loss of motor neurons in the spinal cord. The signals that determine the outcomes of the asymmetric division remain unknown, although various trophic factors are implicated in the proliferation of specific types of neurons, for example, endogenous NGF and other neurotrophins. It is unknown how these processes select some but not other local neurons for death. Many studies show that commitment to cell death is determined after migration, and thus subject to local factors and their stochastic variations.

Despite many fine studies that resolve the scheduling of neuron proliferation, migration, and cell death, our search of the literature found that few reports included numeric presentations of means and standard deviations/errors from which to evaluate the statistics of individual variations in developing brain regions for which the adult individual variations are reported (section 1.4.2). (We do not point out this gap of reported data to criticize many well-designed studies that elucidated the normative developmental patterns; such data are often collected but may not be published because of editorial pressure to conserve space.) In the developing hippocampus and cerebral cortex (Spreafico et al., 1995; Reznikov, 1991), the C-VARs for cell numbers, dividing cells, and dead cells were in the same range (12–60%), with grand means of 25–35% (Tables 3.1, 3.2). Because the C-VARs in the *developing* hippocampus and cerebral cortex brain are also in the same range as those of *adult* brains, we hypothesize that the adult brain variations can be attributed to chance asymmetric divisions that determine cell fates as different types of neurons and neuron survival.

3.2.2. A repeated neuron structure: the lateral line system of frogs

As briefly described for cortical barrels above, neurons can become organized in repeated structures during development. Another example that is understood in exemplary detail is the supraorbital (SO) lateral line

Table 3.2. Interindividual $C\text{-}VAR$ in cell proliferation and cell death in the developing mouse hippocampus

	Pyramidal layer		Dentate gyrus	
Stage	Mitotic cells	Dead cells	Mitotic cells	Dead cells
Embryonic day 20	16%	27%	28%	50%
Postnatal day 1	22%	28%	20%	37%
Postnatal day 21	28%	42%	28%	31%

Mean total mitotic cells: 23.7%
Mean total dead cells: 35.8%

Calculated by C.E.F. from Table 2 of Reznikov (1991).
Mitotic index is the number of anaphases, metaphases, and telophases. Dead cells are the numbers of pyknotic (apoptotic) cells (five mice). The bar graphs of Bayer (1980) and Altman and Bayer (1990a,b) show similar variations in the numbers of ^{3}H-thymidine–labeled neurons.

system of fish and amphibia, which comprises a series of clusters of sensory cells in a repeated structure. Cell numbers in the SO lateral line system of the frog *Xenopus* show striking variations within an individual over a nearly 10-fold range between maximum and minimum (Fig. 3.6A). These variations are consistently observed between individuals and strongly suggest randomness in the determination of cell numbers. Cell number variations in the lateral line arise during development through processes of asymmetric cell division, migration, and cell death as analyzed in elegant experiments (Winklbauer and Hausen, 1983a,b, 1985a,b; Winklbauer, 1988; reviewed in Lewis, 1986; Winklbauer, 1989). The SO system was studied because its cells come from a single primordium and do not incorporate other cells during migration. However, the SO primordium has polyclonal origins, that is, it derives from multiple progenitor cells. This was shown experimentally, by constructing embryos with a mixture of diploid (2N) and triploid (3N) cells (Winklbauer and Hausen, 1985a), which can be identified by cell size. (In general, the size of a particular cell type scales in proportion to the nuclear DNA content.) Lateral line primordia in chimeric embryos were mixtures of 2N and 3N cells, proving a polyclonal origin. Initial thickenings (placodes) on the epidermal cell layer of the embryo head give rise to cells that migrate outward beneath the surface as a column of single cells at a rate of 20–30 µm/hour, which is equivalent to 1–3 cell diameters/hour (Winklbauer and Hausen, 1983a). In Stone's (1993) classic observation:

> The advancing cells have movements . . . that appear jerky . . . with varying resistance to pressure from the . . . migrating primordium. When the migrating primordium has passed, the displaced cells close in again . . . .

Outgrowing neurites follow the migrating primordium, but do not direct its migration; these nerves will eventually connect these sensory

Lateral line development in frog

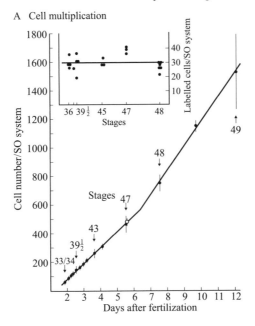

A Cell multiplication

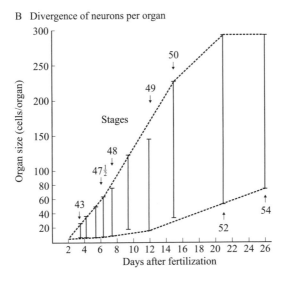

B Divergence of neurons per organ

Figure 3.6. Development of lateral line organs (supraorbital system) in the larval frog *Xenopus*. Larval developmental stages are indicated by arrows. Redrawn from Winklbauer and Hausen (1983b). *A*, Divergence of lateral line organ size (neurons per organ) during development. The range of individual sizes is shown by the vertical lines. The cell numbers reach a maximum by stage 54, whereas the number in the lowest size class remains a relatively constant 6–7 cells. *B*, Cell multiplication, showing cell numbers at each stage; the inset shows DNA labeling by ^{3}H-thymidine. Growth rate is constant: 4.6 cells/hr; cell cycle time of 13.9 hr. *C*, The frequency distribution of organs in each neuron number size class, with modal frequencies of $8 + 7n$. Inset left

(*continued*)

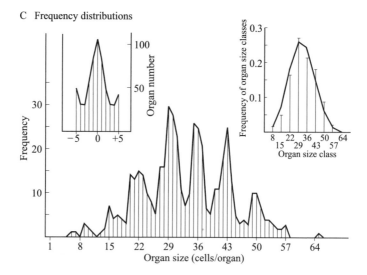

C Frequency distributions

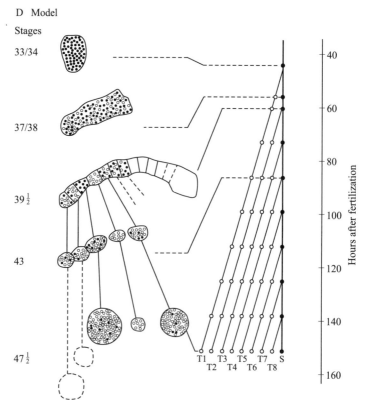

D Model

Figure 3.6 (continued)
shows superposability of the nine peaks; inset right shows organ size classes.
D, Model for the growth and cell proliferation during lateral line development.
Stem cells (solid circles) in the primordium proliferate to produce eight
terminal cells (open circles). The random allocation of stem cells and terminal
cells (postreplicative) in the nascent lateral line organs at stage $39\frac{1}{2}$ is
hypothesized to be the basis for subsequent differences in organ cell number.

cells to the brain. The number of cell clumps left behind during the advance to form the lateral line organs is a characteristic of the species, the location on the body, and the individual's size. Initially, about 16 clumps of cells are formed (16.4 ± 1.5, mean $\pm$ SD). Later in development, another organ is added to the SO lateral line on each side (17.1 ± 1.5). The interanimal variability in the number of organs is one-third the variability in cell numbers *per* organ (*C-VAR*, 9% versus 31%; our calculations).

Cell migration is mediated by filopodia and other membrane specializations, but the molecular mechanisms are not completely known. The migrating stem cells of the *Xenopus* lateral line system express the gene *XkrkI* (Baker et al., 1995). *XkrkI* is related to the proto-oncogene *kit*, a member of the platelet-derived growth factor receptor subfamily, which codes for a receptor that activates tyrosine kinase, an important intracellular signaling step in the regulation of cell division and cell shape.

Variations in cell number per organ become progressively greater during larval development (Fig. 3.6B). By stage 47, the distribution of SO organs sizes comprises nine evenly spaced peaks of cell number classes. The first peak occurs at a median number of eight cells, the next at 15, the next at 22, and so on up to 65, which is the largest peak observed. Each peak is thus separated from its neighbors by seven cells (Fig. 3.6C). The distribution within each peak is approximately symmetrical, as is the overall distribution of the peaks themselves. The highest of the peaks corresponds to the middle cell numbers and the lowest to the numeric extremes.

Further experiments used 3N embryos (Winklbauer and Hausen, 1985a), in which the initial cell clumps contain one-third fewer cells (five or six cells). The resulting distribution of final cell numbers is commensurately reduced, a downsizing of cell number throughout the embryo. The frequency distribution of cell numbers per organ in 3N embryos fitted a binomial distribution from six trials with a probability of success at $P = 0.45$, that is, two fewer rounds of replication than for 2N embryos, but the same chance errors at asymmetric divisions. Adjacent organs rarely exchange cells with each other; otherwise, the smallest size class would be less than observed and more variable within and between 2N or 3N individuals. Nonetheless, the organs are polyclonal, as shown by ploidy chimeras (Winklbauer and Hausen, 1983b). The presence of fewer cells in 3N line organs suggests regulation of organ size, rather than cell number. The reduction in cell number corresponds almost exactly to the increase of cell volume (Winklbauer and Hausen, 1985a). The range of differences between lateral line organs in 2N or 3N embryos is narrowed by about 50%, at a later phase in development when cell division is resumed in the terminal cells of each organ (also see Mohr and Gärner, 1996). The smallest size class increases from 7 to 76 cells; the largest size class increases less, from 64 to about 300. Thus, the final size range decreases by half, which indicates the operation of cell number and pattern correction mechanisms. Other experiments that removed parts of the elongating primordium caused a compensatory extra round of replication; however, fewer organs were formed, each of which had fewer cells than normal (Winklbauer and Hausen, 1985b; Winklbauer, 1988).

Winklbauer and Hausen (1983b) showed that the peak heights themselves (Fig. 3.6C) fitted well to a binomial frequency distribution and proposed the following model to explain the data. Each organ is formed from a clump of eight cells that are left behind as the primordium migrates. The eight cells may be of two types: (i) "stem" cells, which are programmed to have exactly seven asymmetric divisions followed by a final symmetric one, and (ii) nondividing "terminal" cells, which are products of a stem-cell division (the other product being a stem cell, except when the stem cell has used up its division potential and becomes a terminal cell itself). If, when the compartment of eight founder cells is established, each cell independently has a probability of 0.43 of remaining a stem cell and 0.57 of becoming a terminal cell, the model predicts a distribution similar to that which is observed.

When the primordium fragments into cell clumps of about eight cells, the nascent organ is proposed to be a random sample of stem cells and terminal cells. A nascent organ that, by chance, has *only* terminal cells cannot grow further. This numeric class corresponds to the smallest organ found. For the next size class, a nascent organ may be formed from seven terminal cells and one stem cell; then, the stem cell will divide seven times asymmetrically; at each division, one daughter remains as a stem cell while the other becomes a terminal cell. Seven cycles to produce eight new cells result in seven new terminal cells plus one potential stem cell at the last asymmetric division, which does not divide further at this phase; this class of organ, when mature, will have $8 + 7 = 15$ cells, which is the next largest size category observed. The third largest class results when a nascent organ has two stem cells; then 2×7 terminal cells will result, and the organ will have $6 + 16 = 22$ cells. By this process, the final distribution will consist of organ sizes that differ by seven cells, ranging from 8 to 64 cells over an eightfold range (Fig. 3.6D).

Fluctuating asymmetry (section 1.5.3) is also indicated by Winklbauer (1989, p. 201): "Under certain conditions, the left and right SO system grow at different rates, even within the same animal . . . ." It would be interesting to examine how varying degrees of SO asymmetry in larvae may be modified during the large reductions of lateral line systems at metamorphosis. Vital staining with new optic techniques could resolve this question and the related issue of possible asymmetry in the numbers of neurons innervating each lateral line organ.

The model as described above does not account for the dispersion around each peak (it predicts exact numbers in each size class), but there will be several sources of intrinsic and experimental error in the data. Winklbauer and Hausen (1983b) identified the following sources of noise: (1) observer counting errors, which are "small, but not negligible";

(2) chance deviations in cell proliferation, leading, for example, to six or eight asymmetric divisions, instead of seven; (3) chance deviations in the fragmentation of the primordium, so that some initial cell groupings deviate from the median value of eight cells; and (4) the fact that the anterior part of the primordium fragments later than the rest, which would allow continued division by stem cells during migration, and which would reduce the number of divisions available to belated-forming nascent organs. Each of these factors may be important. Moreover, we note that quite different stochastic models may equally well explain the data. For example, each nascent organ may be founded by clumps averaging eight cells that are *all* potential stem cells and become activated by a signal that has a threshold effect, such that the probability of activation for each of them is around 0.43. Whichever model is ultimately correct, the essential feature of the data remains: chance has a major role in determining the final pattern of organ sizes in the lateral line of an individual.

3.2.3. Chance and clonality in intestinal crypts

We now consider a different type of organ, the gut, in which active cell renewal is an essential aspect of function throughout life. The intestinal epithelium is characterized both by a high level of cell production and turnover and also by a well-defined tissue architecture. These properties make it an excellent model to study cell proliferative hierarchies and cell fate (Potten, 1998). The epithelium is organized into hierarchical cell lineages derived from relatively few undifferentiated stem cells that are located near the base of the flask-shaped structures known as the crypts of Lieberkühn. The immediate daughter cells formed by stem cell division acquire stem cell functions under certain conditions, the "potential stem cells" or "clonogenic cells" (Potten and Loeffler, 1990; Potten et al., 1997). These produce cells that undergo rapid clonal expansion, migrating outward and differentiating into mature cells on the surface of the intestinal villi (finger-like projections from the intestinal wall) where they are eventually shed into the lumen of the gut (Fig. 3.7).

This well-defined tissue architecture shows the migration pathways for individual cells. Thymidine-labeling studies define a linear progression down the crypt in cell divisions: cells in the upper regions of the crypt divide once, those slightly lower appear to divide twice, cells farther down three times, and so on. Cells exit the crypt and then move up the villus inexorably at a rate of 1–2 cell diameters/hour (Kaur and Potten, 1986). In mice, the transit time from crypt to the tip of the villus is 2–3 days, and the rate of cell loss from the villus tip is about 1400 cells per villus per day. Most cells in the crypt are replaced every 2 days, whereas the stem cells themselves divide about once a day. During the 3-year maximum life of a laboratory mouse, individual stem cells may divide about 1000 times.

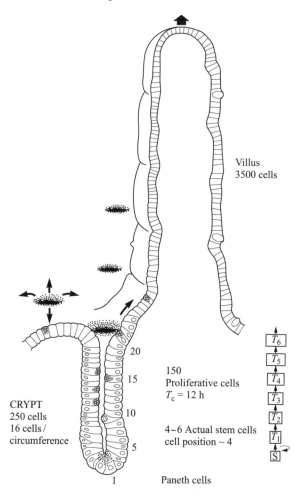

Cell organization of intestinal wall

Figure 3.7. Diagram showing the three-dimensional architecture, sectional profile, and presumed cell lineage organization for small intestine. The position of a cell in the lineage can be related to position along the crypt axis. From Potten (1998), with permission.

Studies of regeneration after injury suggest that the stem cell compartment has a hierarchy of three tiers (Potten and Hendry, 1995; Potten, 1998). At the apex of the hierarchy are four to six ultimate stem cells per crypt, which are highly sensitive to damage, for example, by low-dose ionizing radiation. After exposure to cytotoxins, these stem cells die by apoptosis, which probably serves to maintain the integrity of the stem cell pool. If all of the ultimate stem cells are killed, their immediate daughters (about six in number), which are much more radioresistant and appear to have good repair capacity, seem to assume stem cell functions and maintain the crypt. After even higher dose radiation, this function can be assumed by the next tier of cell progeny (about 24 in number), which again can regenerate the earlier stem cell tiers, the crypt, and the epithelium. This makes a total of about 36 stem and clonogenic cells per crypt. The remaining proliferating cells in the crypt (about 120) have undergone more divisions since they differentiated from the ultimate stem cells and appear to have no clonogenic properties.

Clonal analysis of individual cells in the intestinal epithelium was enabled by a mutation-induced histochemical marker in mice heterozygous at the *Dlb-1* locus, which determines the expression of binding sites for the DBA lectin (Winton et al., 1988). C57BL/6J mice, which express DBA binding sites on intestinal epithelium but not on vascular epithelium, were crossed with SWR mice, which show the reverse pattern. DBA binding is co-dominant, and in F_1 mice both tissues express the binding site. The F_1 mice have only one allele specifying the expression of the DBA binding site on intestinal epithelium, however, and loss of this allele by mutation or somatic recombination in a progenitor or stem cell leads to a clone of cells that do not express the DBA binding site. Mutation in an intestinal crypt stem cell would be expected to give rise to a ribbon of DBA-negative cells extending up a villus, whereas mutation occurring before the crypts form, immediately after birth, should give rise to clusters of wholly DBA-negative crypts in the adult animal. Both patterns were detected in untreated F_1 mice and at a higher frequency in mice treated with the mutagen ethylnitrosourea (ENU). Their absence in homozygotes and their dose dependence on ENU strongly indicate clonal patterns reflecting somatic mutation.

Winton and Ponder (1990) observed the spontaneous accumulation of individual, wholly DBA-negative intestinal crypts during adult life, and their faster appearance following ENU treatment. In the first 4–6 weeks after ENU treatment, more crypts were found that were "segmented" (i.e., partly DBA negative) than wholly negative. But over a period from 6 to 43 weeks the numbers of segmented crypts declined and the numbers of wholly negative crypts increased. Loeffler et al. (1997) have interpreted these findings in terms of a stochastic model of epithelial stem cell organization in which the status of the stem cell population is modeled as a stochastic branching process, where each stem cell produces either two, one, or zero stem cell offspring. Such a model predicts that over time each crypt has an increasing probability of becoming monoclonal, in the sense that the descendants of all but one of the original stem cells have proceeded, by random walk, to become extinct. Furthermore, the model allowed for the possibility of crypt fission, which is thought to occur if the number of stem cells by chance exceeds some threshold number. Li et al. (1994) using DBA F_1 heterozygote mice showed that after ENU treatment a significant number of wholly DBA-negative crypts were found as neighboring pairs, or even clusters. The data on conversion of crypts to monoclonality could not be explained in terms of a model where the key probabilities for a stem cell to produce two, one, or zero offspring were kept constant; some degree of state dependence was required, such that stem cell increase tended to be favored if stem cell numbers became low and suppressed if the number of stem cells approached the crypt fission threshold (Loeffler et al., 1997). That is, the "state-dependence" recognizes some degree of feedback, although the present model does not propose detailed mechanisms for generating such feedback. This stochastic model allowed for (i) state-dependent probabilities of symmetric or asymmetric stem cell division, (ii) stem cell deletion, (iii) effects of ENU on stem cell mutation, (iv) variable cell cycle time, and (v) crypt fission. It provided a comprehensive fit to the available data. Taken together, the experiments and models show a consistent and major role of chance in the intestinal epithelial clonality.

During aging in mice, significant cellular changes occur in the intestinal epithelium (Martin et al., 1998a; Ellsworth and Schimke, 1990). The density of villi and crypts declined by 12–15 months and became progressively sparser. The changes were more obvious in the distal (appendix) region than in the proximal (stomach) region (Fig. 3.8). To examine whether these changes were related to functional alterations in stem cells with age, response to ionizing radiation was measured as a function of age (Martin et al., 1998a). After 1 Gy irradiation, apoptosis was nearly twofold higher in the oldest (29 months) mice versus young and middle-aged groups (Fig. 3.9). After 8 Gy irradiation the levels of apoptosis were uniformly elevated, suggesting that any intrinsic effect of age might have been swamped by the higher level of radiation-induced damage at this dose level. Even without radiation, old rats showed higher cell death and more metaphase chromosome abnormalities in jejunal crypts, with most of the nuclear anomalies in stem cell positions (Ellsworth and Schimke, 1990). These findings are consistent with accumulation of stochastic damage in the stem cell pool. Despite the quality assurance mechanisms acting to prevent accumulation of DNA damage, it seems likely that some damage does accumulate and brings the stem cells in old mice closer to the (low) threshold level required to trigger cell death.

Effects of aging on the capacity of intestinal stem cells to regenerate the tissue after injury by high-dose irradiation were measured in mice (Martin et al., 1998b). The fate of a crypt is determined by whether any stem cells or clonogenic cells repair damage and survive, survival being inversely dose dependent. Sterilized crypts (no surviving clonogenic cells) shrink in size due to cell death and migration of the remaining cells onto the villi. Crypts containing surviving clonogenic cells reform a structure resembling a crypt, with a lumen, Paneth cells at the base, and numerous dividing cells that first repopulate the crypt and then migrate onto the villus. Crypt regeneration processes can be studied using a microcolony assay (Withers and Elkind, 1970). Surviving crypts are

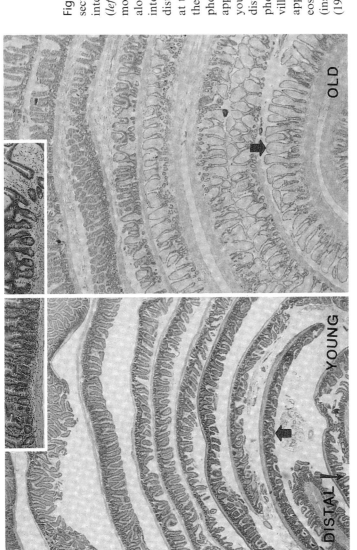

Figure 3.8. Photographs of sections through spirals of small intestine from a 5-month-old (*left*) and 30-month-old (*right*) mouse. The intestine was cut along its length and then rolled into a spiral, beginning at the distal (appendix) end and finishing at the proximal (stomach) end. In the proximal region (upper part of photographs) the crypts and villi appear reasonably similar in the young and old mice, while in the distal region (lower part of photographs) in the old mice the villi have an altered, swollen appearance. Hematoxylin and eosin stain. Magnification, 25× (insets 50×). From Martin et al. (1998), with permission.

Radiation-induced apoptosis in intestinal stem cells

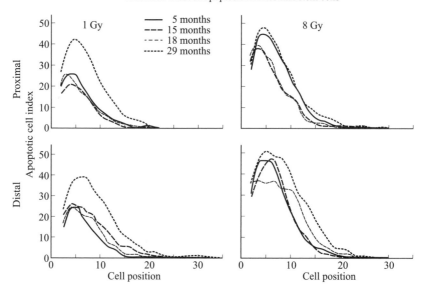

Figure 3.9. Graphs showing the changing apoptotic index at each cell position in the crypt for groups of at least four mice from each of four age groups. The base of the crypt is at the left, with the stem cells at about cell position 4–5, and the top of the crypt to the right at about cell position 24–25. Each graph represents the mean values obtained for 200 half-crypt sections (one side of the crypt) smoothed using a running average over three cell positions. After 4.5 hours of 1 Gy γ irradiation, high levels of apoptosis are observed particularly in the stem cell positions in the oldest mice. After 8 Gy irradiation, apoptotic index is similar in all age groups. From Martin et al. (1998b), with permission.

detected and counted at a time when sterilized crypts will have disappeared.

Crypt survival and growth rate of surviving crypts were measured from day 3 to day 5 postirradiation, and the numbers of clonogens per crypt were estimated from dose–response curves obtained using ranges of radiation doses between 7.5 and 16 Gy (Hendry et al., 1992; Roberts et al., 1995). The surviving crypts in 30-month old mice were both smaller and fewer in number than in 6-month young adults. The growth of surviving crypts, determined by measuring crypt area and number of cells/crypt at various times after irradiation, was delayed by between 0.5 and 1.0 days in the older mice. The number of clonogenic cells per crypt was estimated to be *greater* in the old mice, consistent with Potten's (1992) suggestion that cell death (apoptosis) might stimulate the recruitment of clonogens. Altered levels of apoptosis in the old mice might account for the elevated number of clonogens. However, it may be that the extra recruited cells in the old animals are less efficient clonogens, explaining the delayed regenerative response. An expanded proliferative zone was observed in crypts in the large intestine of old rats (Holt and Yeh, 1989) and humans (Deshner et al., 1988; Roncucci et al., 1988; Paganelli et al., 1990), which is compatible with this suggestion. Refeeding of calorie-restricted rats restored cellular responses in old rats, although there was a broadening of the proliferative zone within the old animals (Holt et al., 1988). These studies suggest age changes in the functions of intestinal stem cells that may have major relevance to individual variations in gut functions during aging.

3.2.4. Hematopoiesis

All of the blood cells in the adult mammal are derived from a population of pluripotent hematopoietic stem cells, which are located in the bone marrow. These stem cells are self-renewing and also give rise to the progenitor cells that become irreversibly committed to one of the hematopoietic lineages. The terminal cell types are mostly short-lived, so the system as a whole relies on a process of continual replacement and renewal. Thus, the burden of maintaining proliferative homeostasis rests with the hematopoietic stem cells. For readers who are not stem cell biologists, it is pertinent that both the gut epithelium system discussed above and hematopoietic system have stem cells which share two important general properties: the continued capacity to proliferate during adult life (unlike mammalian primordial germ cells, section 2.4) and the importance of the local microenvironment and cell–cell interactions in determining the pathways of stem cell differentiation. (see Note, p. 156).

Cell differentiation within the hematopoietic system is regulated by a hierarchy of at least 20 transcription factors, whose overlapping patterns of expression specify the various cell lineages (Enver and Greaves, 1998; Sieweke and Graf, 1998; Rothenberg et al., 1999). Various extracellular growth factors (colony-stimulating factors or hematopoietic growth factors), affect cell proliferation in positive and negative ways. As in other systems, cell fate in the hematopoietic system is determined in a combinatorial fashion at the level of transcription. Several of the transcription factors involved in hematopoietic differentiation are also expressed in other cell types, illustrating the general combinatorial property of cell fate determination.

Multiple transitions and checkpoints are being elucidated in the ten major outcomes of differentiation during hemopoiesis, in which a particular transcription factor or combination of factors need not be lineage specific (Rothenberg et al., 1999). Enver and Greaves (1998) draw on the McAdams and Arkin (1997) model (Fig. 1.16) in arguing that lineage choice results from a probabilistic approach to a threshold at the transcriptional level, possibly arising through stochastic variations in the expression of combinations of transcription factors that determine lineages. Furthermore, Sieweke and Graf (1998) postulate that individual transcription factors are successively replaced in these multimeric complexes, which would allow for various outcomes of gene expression as driven by stochastic fluctuations.

Other stochastic outcomes of gene expression are found in T cell lineages, which present combinatorial assortments of different active cytokine genes, in which the two alleles of a particular cytokine gene are expressed independently (Bix and Locksley 1998; Rivière et al., 1998). These stochastic effects may be adaptive because variations in the production of cytokines by T cells, as selected from a combinatorial inventory, can modify the outcomes of immune responses to infectious agents (Rivière et al., 1998).

Pioneering work by Till et al. (1964) provided early evidence that the pattern of stem cell proliferation is intrinsically stochastic. These studies employed the then newly developed mouse spleen colony-forming assay (Till and McCulloch, 1961), which provides a direct quantitative test for stem cells, based on the observation that mouse hematopoietic tissue contains a class of cells capable of giving rise to macroscopic colonies in the spleens of irradiated animals. The colonies, which form within 10 days, contain in excess of 10^6 cells comprising large numbers of histologically recognizable differentiated cells. Colonies are also capable of self-renewal, since cells that are themselves capable of forming spleen colonies are present within them. Till et al. (1964) dissected individual colonies free of spleen, dispersed their cells, and tested each colony for its own content of colony-forming cells by counting the number of spleen colonies that developed in the spleens of irradiated mice injected with the cell suspension derived from the colony. The actual numbers of cells forming spleen colonies (colony-forming units, CFUs) were multiplied by a fixed factor (5.88) to correct for the fact that only a fraction of potential colony-forming cells injected reach the spleen and there form colonies. By this procedure, Till et al. (1964) obtained the frequency distribution of the estimated numbers of hematopoietic stem cells found in individual colonies. The mean and variance in the numbers of CFUs per colony were 4.5 and 81.4, respectively (*C-VAR* of 200%). Till et al. (1964) observed that the form of the distribution was closely approximated by a gamma distribution (long-tailed) but not by a Poisson distribution, leading them to propose that the number of stem cells in each colony was determined by a random "birth and death" process in which each stem cell would independently and with fixed probability give rise to two stem cell daughters (i.e., an increase by one, or a "birth") or to an early differentiated cell (i.e., a decrease by one, or a "death"). Till et al. (1964) supported their argument with Monte Carlo simulation by computer to demonstrate that the birth and death model did indeed predict a gamma distribution of colony-forming cells per colony.

More recently, Abkowitz et al. (1996) used an alternative approach to demonstrate that hematopoiesis is likely, at root, to be stochastic. They examined female Safari cats, which are F_1 hybrids between domestic cats (of Eurasian origin) and Geoffroy cats (South American wildcats). The two parental cat species are distinguishable by an X-chromosome-linked enzyme glucose-6-phosphate dehydrogenase (G6PD). Because of X-chromosome inactivation during embryogenesis, female Safari cats have some somatic cells that contain domestic-type G6PD (d-G6PD) and others that contain Geoffroy-type G6PD (G-G6PD). On average, the numbers of cells of each parental phenotype were the same; in individual cats the ratio of G6PD phenotypes among committed erythroid progenitor cells equaled that of committed granulocytic progenitors.

In normal circumstances, individual Safari cats showed relatively constant proportions of the two G6PD phenotypes among hematopoietic progenitor cells over extended periods of time (up to 6 years), demonstrating that hematopoiesis appeared polyclonal and stable. However, when six animals were irradiated to destroy their bone marrow and their hematopoiesis was reconstituted by a small number of (1–$2 \times 10^7/$ kg) of autologous nucleated marrow cells, the G6PD phenotypes of progenitors varied extensively, and variation continued for 1–4.5 years after transplantation (Fig. 3.10). These data imply that, although the reduced clonality of the transplanted hematopoietic systems did not fundamentally impair sustained hematopoiesis (at least up to 4.5 years), the effects of stochastic fluctuations were greatly expanded. Abkowitz et al. (1996) used computer simulation of a model in which all stem cell "decisions" (i.e., replication, apoptosis, differentiation/maturation) were determined by chance to demonstrate that stochastic differentiation of stem cells could result in the wide spectrum of outcomes observed in vivo, and that clonal dominance leading to a preponderance of one or other parental phenotype could readily arise by chance. These findings are consistent with early studies of cultured human fibroblasts (Zavala et al., 1977; section 3.4) and of serial transplantations of mouse bone marrow (see below), and may relate to the general observation of age-related reductions in clonal diversity in the immune system (Seldin et al., 1987; Le Maoult et al., 1997).

X-chromosome inactivation patterns have also been studied during aging in women and have revealed evidence of acquired skewing suggesting stochastic clonal loss in hematopoietic stem cells (Gale et al., 1997; Tonon et al., 1998). In peripheral blood of hematologically normal women in three different age groups (17–50, 51–74, and 75+ years), Gale et al. (1997) found a progressive tendency with increasing age for the allele frequencies at X-linked heterozygous gene loci to deviate away from the 50% value predicted on the basis of random X-inactivation. Allele frequencies tended toward either higher or lower values with equal frequency. Such a tendency can be explained by a model that postulates stochastic loss of progenitor stem cells, as further discussed in section 3.4 below.

To ascertain how many clones are active in the hematopoietic system and whether hematopoietic stem cells have indefinite powers of self-

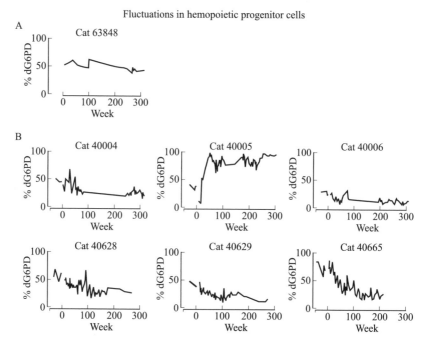

Figure 3.10. The percentage of hematopoietic progenitor cells with d-G6PD genotype as a function of time in individual Safari cats, which are heterozygous for two electrophoretically distinguishable alleles of the X-chromosome linked enzyme G6PD (glucose-6-phosphate dehydrogenase). *A*, Data from a control animal. *B*, Data from six female Safari cats that were irradiated to destroy their original hematopoietic stem cells and in which hematopoiesis was then reconstituted by a small number of autologous nucleated marrow cells. From Abkowitz et al. (1996), with permission.

renewal or are limited in their replicative life spans, Ogden and Micklem (1976) repeatedly transferred hematopoietic cells through a series of irradiated recipients in which the original bone marrow had been destroyed. At each stage a number of stem cells, corresponding typically to 2–5% of the total present in an animal, were injected intravenously. These multiplied and regenerated normal hematopoiesis, and later their descendants were injected into further recipients to repeat the process, and so on. This procedure could be repeated no more than five to six times before the population expired. Ogden and Micklem inferred that there was a progressive decrease in the number of clones that could take part in the repopulation of the hematopoietic system during the course of successive serial transfers. Moreover, serial transfer studies with cells from donors of different ages showed no effect of donor age up to the mean life span (Harrison, 1984). In a series of up to 5 transfers lasting up to 100 months, some host mice maintained normal hemopoiesis, which is several-fold in excess of this species' normal life span (Harrison, 1979; Harrison et al., 1989). Although there are few indications that hemopoietic stem cells in mice undergo significant depletion during aging, the genotype influences stem cell proliferative potential in aging donors (Chen et al., 1999).

As discussed by Micklem and Ross (1978), the picture of an initially large and subsequently declining number of hematopoietic stem cells was open to different interpretations, which have still not been adequately resolved. The decline in clone numbers with successive transfers, and final annihilation of the transferred population, may be a reflection of an aging process that results from cell differentiation fate assignment during development whereby stem cells have only limited powers of self-renewal and proliferation. The repeated stress in repopulating the hematopoietic system, according to this view, merely accelerates the aging process by causing the population to use up its reserves of proliferative potential more quickly than it would under steady state conditions. A contrary view, however, is that the decline is an artifact of serial transfer, possibly resulting in the progressive dilution and eventual loss of the most potent stem cells. These cells may be more difficult to transfer in a viable state or may depend more critically on specific micro-anatomical siting within the bone marrow. The fact that stem cells can sustain hematopoiesis through five serial transfers during 100 months in vivo, which is about 2.5 times the usual mouse life span (Harrison, 1978), shows the ample reserve capacity in the system.

Nonetheless, although hemopoietic stem cells do not appear to become depleted during aging, they do exhibit changes (Morrison et al., 1996). Fluorescence-activated cell sorting was used to separate three categories of multipotent hematopoietic progenitor populations from bone marrow of mice aged 2, 5, 14, and 24 months (see Tables 3.3 and 3.4 for details). Category 1 comprised an essentially pure population of hematopoietic stem cells (long-term self-renewing multipotent progenitors), only 3–4% of which were actively cycling at any one time in 2-month-old animals. Categories 2 and 3 comprised mainly transiently self-renewing multipotent progenitors. Around 7% of Category 2 cells were actively cycling,

Table 3.3. Properties of progenitor cells: three categories of multipotent hematopoietic progenitor populations

Category	Description
1. Thy-1^{lo}Sca-1^{hi}Lin$^-$Mac-1$^-$CD4$^-$c-kit$^+$	Long-term self-renewing multipotent progenitors
2. Thy-1^{lo}Sca-1^{hi}Lin$^-$Mac-1^{lo}CD4$^-$	Transiently self-renewing multipotent progenitors
3. Thy-1^{lo}Sca-1^{hi}Mac-1^{lo}CD4^{lo}	Expanded pool of cycling progenitors

Progenitor populations were separated by fluorescence-activated cell sorting from bone marrow of C57BL-Thy-1.1 mice aged 2, 5, 14, and 24 months. From Morrison et al. (1996).

Table 3.4. Properties of progenitor cells: frequencies of progenitors in the three categories in mice of different ages

Category[a]	Frequencies (% of whole bone marrow cells, mean ±SD)			
	2 months	5 months	14 months	24 months
1	0.007 ± 0.001	0.04 ± 0.01	0.03 ± 0.02	0.05 ± 0.02
2	0.01 ± 0.006	0.005 ± 0.002	0.004 ± 0.002	0.008 ± 0.003
3	0.03 ± 0.006	0.01 ± 0.004	< 0.01	< 0.01

[a]See Table 3.3 for description of categories. From Morrison et al. (1996).

whereas Category 3 was an expanded pool of cycling progenitors of which 18% were actively cycling. As Table 3.4 shows, Category 1 expanded with age while Category 2 remained constant and Category 3 declined, becoming very rare by 24 months.

When functional properties of the Category 1 stem cells were examined, it was found that cells from old (24 month) mice showed a fourfold impairment in engraftment rate of irradiated recipients, which was apparently due to a deficit in cell homing ability rather than to any change in the ability to respond to trophic factors. It was also found that the percentage of actively cycling Category 1 cells was increased about fourfold in old mice. Morrison et al. (1996) suggest that activating hematopoietic stem cells may compensate for hematopoietic deficits that develop in old mice. However, they also point out that driving more of these cells into cycle involves inducing genes or activating proteins that promote cell proliferation and repressing or inhibiting agents that keep cells out of cycle. This runs the risk of increasing the chance of neoplastic transformation by reducing the barriers to proliferation that must be overcome by mutations.

3.2.5. Colony size variation in cell cultures

Hayflick's and Moorhead's (1961) discovery that human diploid fibroblasts have finite replicative capacity in vitro has come to be a major model for cell senescence, although its relationship to cellular changes in vivo remains unclear. The phenomenon of the "Hayflick limit" refers specifically to the number of population doublings (PDs) during serial transfer in vitro. Outsiders to cellular gerontology should be aware that the "Hayflick limit" is not a precise number, because, as we will describe, individual cells vary greatly in their numbers of replications or their proliferative potential. The C-VAR for population doubling of WI-38 fibroblasts from human neonates is 14.6% (data from L. Hayflick, 1968–1972, analyzed by Holliday et al., 1977).

Inverse correlations are reported between the in vitro PD potential during serial culture and donor age (Martin et al., 1970; Schneider and Mitsui, 1976). The largest effect of postnatal age is the twofold greater proliferative capacity of fibroblasts from neonates versus adults (Hayflick and Moorhead, 1961; Martin et al., 1970). However, an extensive study focusing on healthy adults from the Baltimore Longitudinal Study did not detect any change in the PD potential of skin fibroblast lines across the *adult* life span (Cristofalo et al., 1998). The absence of an age trend was confirmed with longitudinal samples from six subjects at ages that were separated by up to 15 years between successive sampling. However, potential age relationships are difficult to resolve because of the major individual variations in the measures of PD potential, which span a five-fold range in proliferative potential across all ages (Martin et al., 1970; Cristofalo et al., 1998). Genetic factors can also influence the degree of variations, for example, the proliferative potentials of fibroblasts cultured from MZ twins had less divergent Hayflick limits than those from DZ twins (Ryan et al., 1981).

The relationship of clonal senescence to aging in vivo is unclear. A comparative study of fibroblast growth from human and seven species of mammals showed a positive correlation between species life spans and their Hayflick limits (Röhme, 1981). Other support for a relationship between in vivo cell aging includes a biochemical marker for senescent cells in vitro that detects some senescent fibroblasts in aged human dermis (Dimri et al., 1995). The limited proliferative capacity of some T-lym-phocyte lineages may also indicate operation of a Hayflick limit (Effros, 1998). However, much as in vivo, when cultured primary cells cease dividing, their postreplicative life span can be considerable. Thus, clonal or proliferative senescence in the Hayflick limit is distinct from, and should not be equated with, cell death or apoptosis (Finch, 1990, p. 422). Replicative senescence in some somatic cells may be a mechanism in tumor suppression (Smith and Pereira-Smith, 1996; Campisi, 1997).

The extensive variations in proliferative potential between cultures of individuals are not a technical artifact. Smith and Hayflick (1974) reported large variations in the PD potential of fibroblast clones isolated from a single mass culture from one individual explant. Moreover, if subclones were grown from individual cells within a *single* clonal popula-tion, the subclones rapidly developed as much variation as the clones sampled from a mass culture (Smith and Whitney, 1980; Fig. 3.11). In these experiments, fragments of pulverized glass with single cells attached were transferred to $1 \, cm^2$ wells to initiate the growth of clones and sub-clones. The cultures contained cells with a bimodal distribution of indi-vidual PD potentials. A subpopulation of cells have low PD potential (< 10 doublings) and did not yield large clones. This subpopulation increased progressively at higher PD levels of the mass culture until, as the mass population approached replicative senescence, all cells had low proliferative potential on subcloning. Another subpopulation comprised

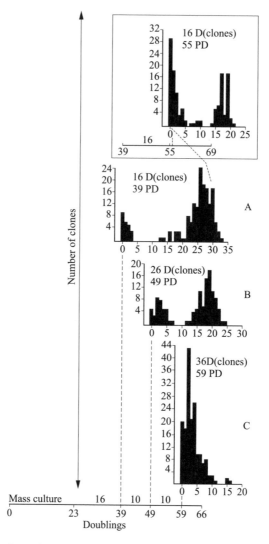

Figure 3.11. Intraclonal variation in cell doubling potential within a culture of human diploid fibroblasts. *A–C* show the doubling potentials of individual cell clones grown from single cells isolated at population doubling levels 39, 49, and 59 from the mass culture. *Top*, The doubling potentials of subclones isolated from a single clone of the PDL 39 series after this clone had itself grown through 16 population doublings from the time it was initiated. From Smith and Whitney (1980), with permission.

cells having a larger (but variable) proliferative capacity, including some whose division capacity approached that of the mass population. By arithmetical reasoning, it is these cells which contribute most to the total cell production in the mass population. When the first series of clones was established in this study by Smith and Whitney (1980), the mass culture had attained a PD level of 39. One clone was used after 16 cell doublings (PD level of 55) to establish a series of subclones, which exhibited a wide range of division potentials, from 0 to 21 PDs. The variability in the PD potential of the subclones was greater than that expected from the variable interdivision time observed for these cells. Smith and Whitney (1980) also found great variability between sister cells of a pair resulting from single mitoses. For example, the proliferative potential of sister cells regularly differed by up to 8 PD, that is, a 256-fold difference in cell number.

Similar asymmetric divisions in hematopoietic cells are also widely seen under culture conditions that minimize microenvironmental heterogeneity (Mayani et al., 1993; Suda et al., 1984; Tsuji and Nakahata, 1989). Cultures from bone marrow or placental cord blood contain stem cells that rise to diverse blood cell types. Sister cells were separated by micromanipulation and placed into separate culture wells. In 10–30% of the pairs, each sister cell gave rise to colonies with completely different blood cell types (erythroid versus granulocyte/macrophage). There was also variation in the numbers of divisions that a cell can undergo at each branch point, which allows for expansion of multipotential lineages.

Human brain glia show similar colony size variations during clonal senescence in vitro (Pontén et al., 1983; Blomquist et al., 1980). Microscopic palladium "islands" were precipitated on a grid on agarose, such that cells grew only on the palladium islands and not on the agarose (Pontén and Stolt, 1980). By choice of cell number when the cultures were seeded, a large proportion of the palladium islands contained one founding cell; in the same dish, a large cell population on the outer circumference of the culture dish was exposed to the media environment, so that media conditioning would occur as in conventional bulk culture. Cells in the miniclones settled and multiplied as efficiently as cells in a mass culture, and represented an unbiased sample of the total cell population (Pontén and Stolt, 1980). The palladium island technique permitted very large numbers of colonies to be scored, with examination of each cell and its progeny. The conclusions were similar to those for fibroblasts. In primary cultures from the frontal lobe of a 38-year-old male, all clones of "glia-like" cells gave a broad distribution of colony sizes (Pontén et al., 1983). Even at the earliest stages, cultures had substantial numbers of small colonies, although many single cells yielded colonies of 16 cells or more (four or more cell doublings) within the 10-day experiment. With increasing culture age, the distribution shifted to fewer large colonies and more small colonies, until at the oldest culture age only small colonies remained, due to the decreased subpopulation of cells capable of proliferation.

In conclusion, substantial stochastic variation in clonal PD potential appears to be intrinsic to diploid cell cultures. Chance variations in cell clones are well recognized, of course, for their potential relationship to pathogenic effects in vivo, such as cancers, skin keratoses, and athero-sclerotic plaques. The asymmetry in proliferative potential between sister cells in a clone is consistent with stochastic features in the regulation of genes that determine the proliferative potential, but the molecular basis is not known. Other views on the stochastic and epigenetic variations of clonal proliferation are discussed by Cristofalo and Pignolo (1996) and by Rubin (1997).

3.3. Somatic genome instability

Among the many stochastic factors that influence cell fate, one of the most important to outcomes of aging is somatic genome instability lead-ing to an extensive degree of somatic genomic mosaicism. There is now ample documentation of somatic mutations in chromosomal and mito-chondrial genomes and chromosomal abnormalities, which comprise a powerful general drift away from the unique genomic identity of the zygote to somatic genomic mosaicism of the aging adult. We focus here on genome instability affecting the nuclear genome; instability affecting the mitochondrial genome was discussed in section 1.7.2. Sources of somatic genome instability range from intrinsic errors in DNA replica-tion through a wide variety of intrinsic and extrinsic factors that may damage not only the genome but also other cell components. Epigenetic modifications, for example, through loss or disruption of DNA methyla-tion patterns, may also be pertinent (Holliday, 1987). Most of these changes will be deleterious and may contribute to senescent phenotypes. Their roles in carcinogenesis are well recognized.

Somatic mutations were the first molecular hypotheses of aging (Failla, 1958; Szilard, 1959; Burnet, 1974; Kirkwood, 1989; Slagboom and Vijg, 1989). Intrinsic mutagenesis, as Burnet (1974) argued, is an inevitable source of random genetic diversity in development (including the adaptive antigen-driven hypermutation in immunoglobulin V–region coding sequences that fine-tune antibody responses), also in carcinogen-esis and senescence. There is clear evidence for an age-related accumula-tion of somatic mutations during aging in human lymphocytes (Morley et al., 1982; Trainor et al., 1984; King et al., 1994). While less explored, there will be also be degrees of spontaneous somatic gene mutation and somatic cell mosaicism at prematurational stages of development. The issue is not whether somatic mutations occur, but whether they occur with sufficient frequency to contribute significantly to chance variation in *functional* outcomes of aging besides those associated with cancer.

To study mutational events in organs and tissues of aging animals directly, Vijg and co-workers (Gossen et al., 1995; Boerrigter et al., 1995; Dollé et al., 1996) developed transgenic mice with bacterial *lacZ* reporter genes within chromosomally integrated plasmids. The bacterial genes serve as "targets" for mutational events, and the frequency of mutations in different tissues and at different ages can be estimated by rescuing the plasmids from mouse tissue and scoring for *lacZ* mutations. This system thus allows positive selection of mutant colonies from as many as a million wild-type cells and can detect low-frequency mutations. Further refinements using polymerase chain reaction (PCR) detect various mutations (Dollé et al., 1997), including point mutations and various size change mutations, with internal deletions and rearrangements that may include the flanking mouse DNA sequences. The frequency of mutational events during postnatal aging differed between brain and liver (Fig. 3.12

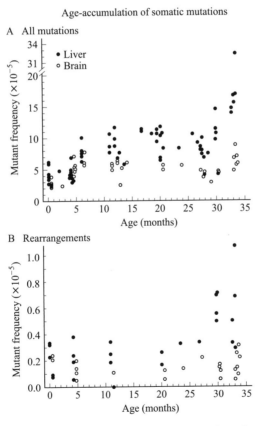

Figure 3.12. Frequencies of mutations as a function of age in transgenic mice carrying bacterial *lacZ* reporter genes. *A*, All mutations; most were point mutations or small size changes; liver accumulated more than brain. *B*, Rearrangements and deletions. From Dollé et al. (1997), with permission.

and Table 3.5). Liver showed a progressive increase with age in frequency of mutations not involving change in the restriction digest pattern (no-size-change mutations) from postnatal day 1 up through senescence (34 months), when the mutant frequency was 12.2 ($\pm$5.8) $\times$ 10^{-5} (Fig. 3.12A). In brain, mutations approximately doubled between infant and young adult ages, but did not increase further at later ages beyond a frequency of about 5 $\times$ 10^{-5}. The frequencies of mutations involving major genome rearrangements (size change mutations) showed some increase with age in liver but not in brain, but were an order of magnitude or more lower than the point (no-size-change) mutations (Fig. 3.13B).

These powerful experiments permit comparisons of mutation frequencies among different organs, but were not designed to resolve exactly which types of cells were involved. The extent of mutations in neurons is of great interest, because most neurons in the adult brain are post-replicative and because of evidence at a neuron cell level for frame-shift mutations that increase progressively during aging in rodent and human brain neurons (van Leeuwen et al., 1998; Finch and Goodman, 1997). Various glial types have low levels of cell turnover in the adult brain, which include occasional cell cycles in the brain, as well as continued influx of bone marrow-derived monocytes into the brain, which repopulate the microglia. Thus, the leveling off of mutation frequency before middle age as measured in bulk brain tissue might reflect a steady state value due to glial and vascular endothelial cell turnover, which would obscure possible unique effects on neurons. The need to consider different cell types applies also to the liver, which contains slowly dividing hepatocytes and tissue macrophages (Kupffer cells) that are repopulated by marrow cells. The measurement of somatic mutation frequencies is

Table 3.5. Summary of mutant frequencies by age group measured in transgenic mice with bacterial *lacZ* reporter genes within chromosomally integrated plasmids

Age groups	Mean mutant frequency ($\pm$ SD, $\times 10^{-5}$)		Mean frequency of major genome rearrangements ($\pm$ SD, $\times 10^{-5}$)	
	Liver	Brain	Liver	Brain
Neonate (1 day)	3.9 $\pm$ 1.1 ($n = 12$)	2.8 $\pm$ 0.5 ($n = 8$)	0.30 $\pm$ 0.05 ($n = 4$)	0.16 $\pm$ 0.08 ($n = 4$)
Young adult (4–6 mo.)	5.4 $\pm$ 2.0 ($n = 18$)	4.8 $\pm$ 1.6 ($n = 15$)	0.21 $\pm$ 0.13 ($n = 4$)	0.12 $\pm$ 0.06 ($n = 4$)
Middle-aged (10–24 mo.)	9.4 $\pm$ 1.8 ($n = 23$)	5.4 $\pm$ 1.1 ($n = 12$)	0.25 $\pm$ 0.07 ($n = 6$)	0.07 $\pm$ 0.06 ($n = 6$)
Senescent (25–34 mo.)	12.2 $\pm$ 5.8 ($n = 19$)	5.0 $\pm$ 1.4 ($n = 17$)	0.53 $\pm$ 0.24 ($n = 11$)	0.14 $\pm$ 0.08 ($n = 11$)

From Dollé et al. (1997).

also complicated by intragenomic heterogeneity (for example, Gossen et al., 1993), which includes effects of transcription on DNA repair rates (e.g., Bohr et al., 1985; Madhani et al., 1986). Nevertheless, the data from Dollé et al. (1997) indicate that frequencies of mutations up to 10^{-4} per gene occur during aging in mice. This conclusion is also supported by studies of the "senescence-accelerated mouse" (SAM), strains that prematurely develop different senescent phenotypes in association with shortened life spans (Higuchi, 1997; Takeda et al., 1997). In the SAMP1 substrain, the mutation frequency at the hypoxanthine phosphoribosyl transferase (*Hprt*) gene locus in splenic lymphocytes increased 100-fold, from about 10^{-5} at 3 months to 10^{-3} by 9 months and older (Fig. 3.13; Odagiri et al., 1998). This compared to an increase in mutation frequency from about 10^{-5} in 3-month-old mice to around 10^{-4} at 12 months in another strain (SAMR1) from the same background, which had normal aging. In SAMP1 mice, the mutation frequency leveled off at 10^{-3} and

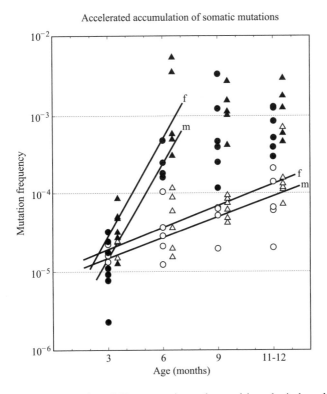

Figure 3.13. Frequencies of *Hprt* mutations observed in splenic lymphocytes from male (m, circles) and female (f, triangles) senescence-accelerated mice (SAM). Solid symbols, strain SAMP1 (rapid senescence; median life span 9.9 months); open symbols, strain SAMR1 (median life span 18.9 months). From Odagiri et al. (1998), with permission.

did not further increase up to the ages examined, suggesting that a muta-tion frequency this high, if experienced at other genetic loci as well as *Hprt*, may be a limit to what the lymphocytes could tolerate.

Assuming 50,000 or more functional genes in mammalian cells, the data from the various studies described above suggest that many, and possibly most, somatic cells are likely to carry at least one chromosomal DNA mutation. In addition, there will be a substantially greater cell burden of mitochondrial DNA mutations which arise at a much faster rate than in the nuclear genome (section 1.7.2).

3.4. Clonal attenuation, cell immortalization, and cancer

We discussed in section 3.2.5 how colony size variations are a feature of cultured cells. These variations have important implications for the clonal behavior of cell populations during aging and also during carcinogenesis. If individual clones have varying division potentials, each individual clone within the population will, at different times, form a varying pro-portion of the total cell population, either in vitro within a culture or in vivo locally within a tissue. If the clone has a finite replicative life span (Hayflick limit), then the clone will be numerically greatest at a time shortly before it is extinguished. This process of clonal expansion and disappearance within mass populations was termed "clonal attenuation" by Martin et al. (1974).

Evidence for clonal attenuation by Zavala et al. (1977) was interpreted in these terms by Holliday et al. (1977). Skin fibroblast cultures were established from women who were heterozygous for glucose-6-phosphate dehydrogenase (G6PD) alleles A and B, which are electrophoretically identified (Zavala et al., 1977). Since G6PD is X-chromosome-linked and only one X-chromosome is active in a somatic cell (inactivation of the other X-chromosome being random), individual cells have either G6PD A or B enzyme phenotypes. Most fibroblast cultures had close to equal numbers of cells with one X inactivated over many population doublings, although in a few cases the ratio of A to B diverged steadily through successive subcultures, which indicates differential growth. Nonetheless, as the replicative life span was approached in many cultures, there was much variability, sometimes converging quickly to one active allele, sometimes oscillating. Similar findings were made during serial marrow transplantation of hematopoietic cells through successive irra-diated hosts (Ogden and Micklem, 1976; see section 3.2.4). The T6 chro-mosome marker, either heterozygous or homozygous, was used to track donor cells. In mixtures of the two cell types, the ratio remained constant for three or four serial transplants but in the last one or two transplant generations, before cells died out, one or other cell type came to predo-minate.

The late-arising fluctuations in proportions of cell phenotypes reported by Ogden and Micklem (1976) and Zavala et al. (1977) are fully consistent with the idea of clonal attenuation and with a stochastic model derived from the "commitment theory" of Kirkwood and Holliday (1975, 1978). This model assumed that cells are initially uncommitted and have indefinite replicative potential, but that, upon division, an uncommitted cell produces committed daughters with some probability (P). Committed daughters are assumed to have finite replicative potential and to generate clones of cells that cease division after M cell generations. There are thus three possible outcomes to the division of an uncommitted cell: two uncommitted daughters, probability $(1 - P)^2$; one committed and one uncommitted daughter, probability $2P(1 - P)$; and two committed daughters, probability P^2. A direct prediction from this model is that a population which may begin as entirely uncommitted will rapidly generate extensive clonal diversity, and furthermore that the replicative potential of the population as a whole will be strongly influenced by what happens to the dwindling fraction of uncommitted cells. Even if the probability of commitment P is < 0.5, so that an uncommitted cell produces on average > 1 uncommitted daughter, the *proportion* of uncommitted cells will decline as the number of committed progeny builds up. According to this model, eventually there may come a time when the last uncommitted cells are lost, and the population then comprises a set of cells that have highly variable division potentials but all of them finite. The loss of the last uncommitted cells is likely by chance, but the randomness that is involved here does not become apparent until other clones are extinguished. As the population growth approaches its end, however, the population comes to be dominated by the fates of those last uncommitted cells. This predicted behavior (Holliday et al., 1977) was observed by Ogden and Micklem (1976) and Zavala et al. (1977). It is also predicted that as the population continues through the earlier stages of growth, individual committed clones are continually expanding and then being extinguished and replaced. This is consistent with observations that cells with little or no division potential are detected quite early in culture life spans and that individual dermal fibroblasts exhibited a marker for cell replicative senescence in vivo long before major deficits in proliferation of the tissue (Dimri et al., 1995).

The key point about the commitment model, and the idea of clonal attenuation, is that although most of the cells that can be observed at any one time come from clones that have already expanded, the long-term growth potential of the population rests with a relatively tiny number of cells that have much greater proliferative potential than the rest. A direct test of this property of cell cultures was a "bottleneck" experiment (Holliday et al., 1977). If it is true that the population contains a few cells with much larger PD potential than the rest, then a drastic reduction in culture size should lead to the premature elimination of these cells, and this will be reflected in a reduced number of PDs for the culture as a whole. Holliday et al. (1977) grew populations of human diploid fibroblasts in normal culture flasks containing about 2×10^6 cells and at each of four PD levels isolated 8–12 bottleneck populations of 2×10^3 cells. The bottleneck populations were allowed to grow back up to normal size and thereafter cultured normally until they underwent replicative senescence. The parental cultures had an average replicative life span of 54 PDs, and bottleneck cultures taken at PD levels of 8, 13, and 21 had their replicative life spans reduced by an average of eight population doublings, that is, a 64-fold reduction in cell production. The data (mean reduction and variance) agree exactly with the predictions from the model (Holliday et al., 1977; see Kirkwood and Holliday, 1978; Fig. 3.14).

A corollary to the prediction that bottlenecking should reduce replicative life span when there are few cells with high doubling potrential is that later bottlenecks should have little or no effect on the "culture life span," if they are taken when the rare cells with high replicative life spans already have disappeared from the mass population. Bottlenecked cultures taken at PD level 31 had negligible effect on the culture life span. The fact that late bottlenecks (PD 31) had only a small effect on the culture life span, whereas earlier bottlenecks (PD 8, 13, and 21) had major effects, confirms that it was not bottlenecking per se that caused the reduction in division potential of the population. It is hard to construct an alternative explanation for these results that does not include the loss of rare cells ($< 10^{-3}$) in early cultures with high proliferative potential and/or self-renewal.

The rationale for testing the commitment model (Kirkwood and Holliday, 1975) was to investigate whether this kind of population process could account for the difference between finite growth, as seen in normal diploid cultures, and indefinite growth, as seen in immortalized cell lines. The model shows that this was indeed possible and that the major qualitative dichotomy between finite and infinite growth depends on the quantitative parameters P (probability of commitment), M (number of cell generations before a committed clone ceases replication), and N (effective population size). If P or M are small enough, then the dilution of uncommitted cells by committed daughter cells is arrested by the fact that the committed daughter cells start to die before the uncommitted cells are lost.

Whether or not normal cell cultures actually contain, at an early stage of their growth, cells capable of unlimited growth, this population process is also relevant to the growth kinetics of immortalized cell lines. It is not necessary that an immortalized cell line contains a majority of cells capable of founding an immortal cell lineage. What matters is that a sufficient fraction of the cells can do this and that the cells which are unable to do so reach their replicative limit fast enough that they do not dilute out those that can. Rafferty (1985) showed that the fraction of cells in a HeLa line that formed permanent subclones was only about 70%. In a remarkable study of immortalization in cultured rat liver epithelial-like cells (Matsumura et al., 1985), time-lapse cinema microscopy showed the

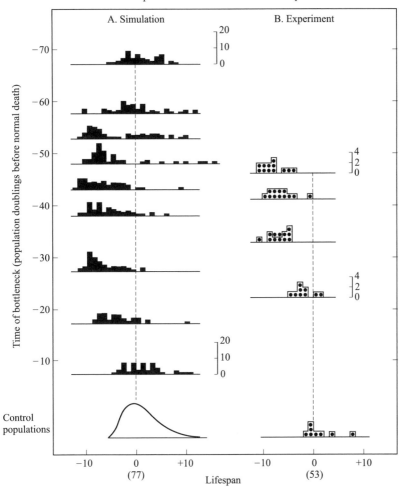

Figure 3.14. Histograms of simulated and experimental human fibroblast culture life spans, subjected to a transient 1000-fold reduction in culture size ("bottleneck") at various times during culture growth. The control populations at the bottom of the diagram were not subjected to a bottleneck. To facilitate comparison between simulated and experimental data, all times and life spans (both measured in terms of cell population doublings) are shown relative to the mean population doubling level at replicative senescence (equated in the original paper to proliferative "death") of the control populations. The simulations are according to a stochastic model for commitment of cells to senescence (Kirkwood and Holliday, 1978; see text). From Holliday et al. (1977), with permission.

gradual emergence of permanently growing subclones from a founder population in which the great majority of cells had finite growth potential (Fig. 3.15). In the early phase of immortalization, one of the component cells changed and gained an additional proliferative potential. In this phase, many cell lineages were still finitely proliferative, while some continued division, although unstably.

A major hypothesis for the finite proliferative capacity of cultured diploid cells (the Hayflick limit) is the observed progressive loss of repeated DNA sequences at chromosome ends (telomeres; section 1.7.1). This progressive DNA erosion is attributed to insufficient expression of the telomerase gene in most mammalian somatic cell types (Thomson et al., 1998; Weng et al., 1998; Harley and Sherwood, 1997). Gradual telomere shortening in the absence of telomerase is subject to chance variations during DNA replication, which could generate progressively greater chromosomal heterogeneity among the individual cells within a mass population. Rubelj and Vondraček (1999) also predicted the occurrence of abrupt telomere shortening, arising through DNA recombination or nuclease digestion at the subtelomeric–telomeric

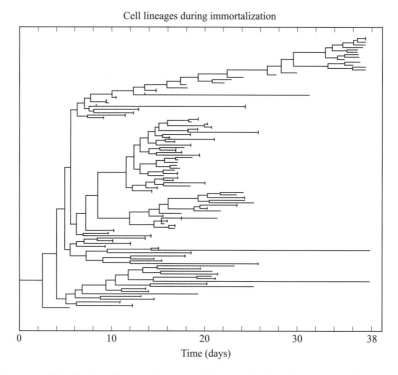

Cell lineages during immortalization

Time (days)

Figure 3.15. The family tree of a rat cell lineage during the process of immortalization, as reconstructed from time-lapse cinematography. From Matsumura et al. (1985), with permission.

border. Their stochastic model of such processes gave predictions that match the pronounced clonal heterogeneity in proliferative potential (Smith and Whitney, 1980; Fig. 3.11).

The fibroblast model was recently developed further with the demonstration that the Hayflick limit can be broken by restoring telomerase activity. In a unique experiment, Bodnar et al. (1998) reintroduced telomerase into diploid human cells (foreskin fibroblasts and retinal pigment epiethelial cells). These engineered cells have exceeded their Hayflick limit by fivefold (they are still growing), but without indications of impaired growth control as is typical of cancer cells. Although many details of this rescue remain to be described, the results suggests that the restoration of telomerase activity overrides the propensity for stochastic loss of telomere DNA during extended replication in cells with low telomerase activity. Also see Counter et al. (1998) and Russo et al. (1998) for other aspects of telomerase immortalization. Thus, modulations in the activity of a single gene may, in some contexts, have major effects on stochastic processes of aging. However, other mechanisms are also at work in the Hayflick limit besides telomere erosion, including activation of tumor suppressor genes (*Rb* and *p53*) and repression of growth stimulatory genes (Smith and Pereira-Smith, 1996; Campisi, 1997; Garkavtsev et al., 1998). Moreover, signaling pathways may connect telomere structures with cell cycle machinery (Harley and Sherwood, 1997). But whatever the roles of these various mechanism(s) may prove to be, they must be consistent with the compelling evidence for stochastic factors in determining replicative potential. Despite the strength of the findings on telomerase immortalization, many other factors may determine the stochastic features of cell proliferative potential. Moreover, mice that totally lack telomerase through a genetically engineered germline deletion have normal life spans and patterns of aging (Rudolph et al., 1998). Only after 3 generations does mortality increase at earlier ages, accompanied by impairments in wound healing, hemopoiesis, intestinal atrophy, lymphomas, and lymphocyte chromosomal abnormalities. The role of telomere loss and stabilization remain to be fully clarified with regard to cell replicative senescence and tumorigenesis. However, the mouse telomerase knockout experiment shows definitively that the absence of telomerase in *all* somatic cells and supranormal telomere DNA loss do not modify aging, which strongly implies that somatic cell telomeric DNA loss is not the major determinant of life span in normal laboratory mice.

3.5. Conclusions

We considered how chance variations affect cell fate and how stochastic developmental effects at the level of individual cells give rise to increasing cellular heterogeneity during aging. Asymmetry in cell fate is intrinsic to much of development. Nonetheless, there are extensive cell-cell interactions through which initially variable patterns of gene expression become restricted. The random-walk Brownian motion variations in RNA and protein synthesis are an intrinsic source of fluctuations, which are thought to be a major mechanism in the assembly of transcription factors during bone marrow stem cell differentiation. Cell migration errors provide additional opportunities for variation in outcomes of development. Programmed cell death, which is essential to early development and to homeostasis of some adult tissues, shows substantial variability within and between individuals.

As multicellular organisms develop, they become, effectively, a large variegated cell clone that contains a mosaic of individual subclones. Stochastic variations are seen in the clonal dispersion of neurons in the developing brain, in the clonal development and homeostasis of intestinal epithelium based on division of stem cells in the crypts of Lieberkühn, in the clonal expansions that form the cells of the hematopoietic lineages, and in colony size variations within populations dividing cells such as fibroblasts and glia. Clones of proliferating cells expand and become attenuated through cell replicative senescence, or may at random become immortalized. The proliferative behaviors of individual cell lineages during cellular immortalization and in cancer cell lines reveal marked, intrinsic heterogeneity.

Instability of the somatic genome includes mutations in the chromosomal and mitochondrial genomes; chromosomal instabilities; and epigenetic modifications affecting DNA methylation patterns. All of these contribute to progressive random variations in somatic cell genomic integrity with aging. In turn, instability of the somatic genome contributes directly to the age-related increase in carcinogenesis, which itself is a stochastic process. The apparent protection of telomere DNA loss by restoration of telomerase in fibroblasts is a precedent for considering the enzymatic-biochemical modulations of other chance molecular damage during aging.

Note

Largely for historical reasons, the terminologies used to describe apparently similar processes by researchers in these subfields of stem cell biology are different. Perhaps for reasons of the prevailing "microculture" within each field, few cross references are found in current publications.

4

Chance in the Developmental Environment

We now consider effects of the *external* environment on *prenatal* development, with an emphasis on aspects of the nervous system already discussed in the context of *intrinsic* chance fluctuations. Our focus causes us to neglect many examples of other systems that are sensitive to external fluctuations. The external environment envelops the embryo, within which arise chance variations at the level of molecules and cells, such as the variations in fetal blood sex steroid levels (section 2.6). We describe how chance variations in environment surrounding the embryo and in the maternal environment modify development, but can also modify the outcomes of aging in adults. Among other examples, a fetus may be subject to (i) different levels of sex steroids according to the chance that its fetal neighbors are male or female, (ii) chance exposure to maternal stress, and (iii) chance exposure to environmental estrogens. Environmental estrogens that modify the adult prostate size and biochemical characteristics in rodent models may influence risks of benign prostatic hyperplasia during aging (section 2.6.2). We also briefly review examples of prenatal nutritional effects that may influence risks of later disease. In rodents, effects of malnutrition can persist several generations.

4.1. Fetal interactions

The levels of sex steroids in the fetal circulation determine which types of external genitals, male or female, develop in humans and other mammals (see Fig. 2.11). The fetal testis produces testosterone and müllerian inhibitory factor, which masculinize other components of the reproductive sytem. Most components of the reproductive system have a critical period during development, when variations in hormone levels (time and dose) will irreversibly modify adult characteristics and, in some examples, outcomes of aging. These *organizing* hormonal effects establish the adult anatomy and physiology organ functions. Steroids or molecules with steroidlike activity from the environment can also pass into the fetal circulation, where they may function as endocrine disruptors. We will consider interactions of sex steroids produced by adjacent fetuses, in which chance variations in the distribution of male and female ova in the uterus lead to shadings of gender with different outcomes during aging.

4.1.1. Rodents

Definitive examples of gender interactions between fetuses were brought to light in a two-decade-long series of remarkable studies led by Fred vom Saal. To help grasp these findings, we briefly review some basic biology. Mice, rats, hamsters, and other rodents typically have multiple births, 5–15, which arise from separately fertilized eggs (nonidentical "tuplets") that are distributed in both horns of their uterus. The number of embryos in each horn of the uterus is an outcome of chance variations in the numbers of unfertilized eggs that enter through the oviduct. The adult ovary has a succession of follicles at different stages of maturation characterized by different numbers of granulosa cell layers surrounding the oocyte (Figs. 2.1, 2.2). Although the number of growing follicles is not tightly regulated, survival of larger follicles depends on follicle-stimulating hormone (FSH), which is secreted by the pituitary under exquisitely sensitive hormonal feedback from the growing mass of follicles (Hirshfield, 1991). Consequently, the number of follicles that ovulate is relatively constrained, although the mean number varies among mouse genotypes. Which of the growing follicles survive and which die through atresia are chance events of growth stage and levels of FSH.

In rodents, the uterus is duplex with two horns (bicornuate). Each uterine horn is connected to the ovary on that side, so unfertilized eggs from each ovary are shed at ovulation into the uterine horn just on one side, where they are fertilized. Then, at fertilization the sex of an individual is determined by a second chance event. Before fertilization, each egg normally carries a single X-chromosome. Half of the sperm carry a Y-chromosome, and the eggs that they fertilize develop as males; the other sperm carry an X-chromosome and the eggs they fertilize develop as females. Thus, an egg is destined by chance at fertilization to become a male or female embryo.

The next chance event is whether a particular fetus has fetal neighbors of the same or opposite sex (vom Saal, 1981; vom Saal et al., 1998a; Fig. 4.1). Thus, each fetus can be classified by the sex of its neighbors. By convention, a male flanked by two males is a 2M male, whereas a female flanked by female neighbors is a 2F female. Other fetal neighbor combinations do not concern us here, but fit into a continuum of gender dose effects. The array of fetal neighbors fits a random distribution in litters with different numbers of pups (Fig. 4.2). For example, the proportion of 2M females in a litter is 1 in 6. Thus, chance has already operated on three different processes before development begins: the growing follicles that survived to ovulate, the numbers of females and males, and the sex of the neighboring fetuses.

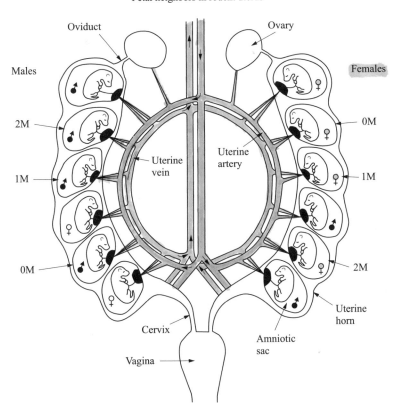

Figure 4.1. Uterine anatomy of rodent showing fetal neighbor designations and blood flow. Symbols are for male and female pups: 0M, flanked by no males; 1M, flanked by one male; 2M, flanked by two males. Redrawn from vom Saal and Dhar (1992).

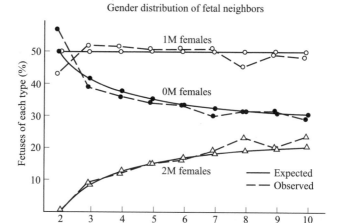

Gender distribution of fetal neighbors

Figure 4.2. Frequencies of 0M, 1M, and 2M female mice observed in 1513 uterine horns, plotted by the dashed line. The designation 1M female can represent a male flanked by one male and one female or, if at either end of the uterus, could also represent a male flanked by only one pup, an inside male, but without contacting any female. The solid line shows the expected distribution if fetal positions were random. The numbers of pups per uterine horn varied from 2 to 10, with a mean of 6 pups. Redrawn from vom Saal (1981).

The random distribution of fetal neighbors of different sexes has major consequences for development, because the blood levels of sex steroids in each fetus are influenced by the sex steroids from neighboring fetuses in mice and other microtine rodents (vom Saal, 1989). In lab mice of either sex at later stages of gestation, the female sex steroid estradiol is 30% higher in fetuses that are flanked by females: 2F males or 2F females (Fig. 4.3A). Reciprocally, testosterone is correspondingly higher in fetuses flanked by males: 2M males or 2M females. These differences are consistent with higher blood estradiol in female fetuses and of testosterone in male fetuses. Thus, both males and females cast steroidal shadows onto their neighbors that cause subtle shadings of gender in neural systems during critical phases of development.

How these differences in steroid levels are transmitted between fetal neighbors is unclear. In mice, each embryo has a separate placenta, which is distinct from its neighbors on each side. Experiments show negligible transmission of sex steroids by blood flow between adjacent fetuses. Direct vascular connections between the adjacent placentas are rare (1%; vom Saal, 1981). Among other possibilities, sex steroids can be passively exchanged between neighboring fetuses across the placental membranes (Even et al., 1992). Another possibility is changes in the rate of fetal blood flow in late gestation, which differs by gender (Even et al., 1994).

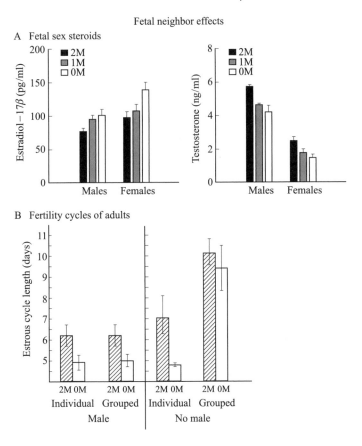

Figure 4.3. Fetal blood sex steroids influence the gender of fetal neighbors. *A*,
Estradiol is 25–35% lower in male or female pups (2M) with males as
neighbors, whereas testosterone is correspondingly greater. Redrawn from vom
Saal (1989). *B*, Fetal neighbor effects on estrous cycle length of adults.
Redrawn from vom Saal and Bronson (1980).

Fetal neighbor effects have major influences on adult reproduction of
mice and other rodents (vom Saal et al., 1994, 1998a; Clark and Galef,
1995). In young adult mice, 2M female mice have longer and less regular
estrous cycles (intervals between ovulation) than 2F females (Fig. 4.3B).
Even so, the numbers of live-born pups and their birth weights do not
differ between 2F and 2M females in controlled mating studies. These
cycle length phenotypes, having originated by chance in the uterine environment, are also subject to chance features in the adult social environment: notice that cycle length no longer differs between 0M and 2M
females when females are placed in groups without a male (Fig. 4.3B).

Fetal neighbor effects modify adult behavior. The behavior of a
female influences the success of her mating, depending on environmental

factors such as population density and intrusion by other rodents. The 2M females show more aggression to conspecific intruders and other masculine behaviors, such as tail rattling and chasing and biting of 2F females. Thus, 2M females have a reproductive advantage over 2F females at higher population densities. However, this enhanced territoriality, also seen in 2M males, because of the wounds incurred during fighting, would be predicted to increase mortality risk and shorten life expectancy.

Similar fetal neighbor effects on adult behaviors are also observed in wild mice (Palanza et al., 1995), which lowers concerns about artifacts of laboratory inbreeding. Gerbils and rats also show fetal neighbor effects, although these species differ in the traits modified, for example, in parental effort and infanticide (vom Saal et al., 1994, 1998a; Clark et al., 1991). Fetal neighbor effects may also extend to pigs, because females from litters with a preponderance of males had lower fertility (Drickamer et al., 1997; vom Saal et al., 1998a). Sheep may also have similar fetal neighbor effects on fertility due to higher fetal mortality (Avdi and Driancourt, 1997). If fetal neighbor effects are borne out for commercialized animals, manipulations of the sex ratio could be a target for enhanced fertility. The species variations indicate that fetal neighbor effects are under selection in response to population density and the local environment.

These fetal neighbor variations are epigenetic, being mediated by quantitative effects of hormone levels on cell signaling systems. Moreover, there is yet another multigenerational epigenetic impact: the offspring of 2M female gerbils and mice themselves give birth to litters with about 35% more sons (Clark and Galef, 1994; Vandenbergh and Huggett, 1994; vom Saal et al., 1998a). These effects extend to at least the second litter in mice (Vandenbergh and Huggett, 1994). In gerbils, infanticide is not the factor by which the fetal neighbor influences the sex ratio of the next generation (Clark and Galef, 1995). The variations in sex ratio are among many other observations showing sensitivity of the sex ratio in response to the environment (Trivers and Willard, 1973; Meikle and Thornton, 1985; Vandenbergh and Huggett, 1994; Kilner, 1998). The mechanisms mediating these phenomena are not known but clearly involve neuroendocrine systems at some level. They present a valuable example of the transgenerational persistance of an epigenetically acquired trait.

Fetal neighbors also influence outcomes of aging in both male and female mice and gerbils (vom Saal et al., 1994, 1998a). Again, the pattern is complex. The first few litters were the same size in all neighbor categories. However, during aging, 2F females became infertile about 3 months before 0F females (Fig. 4.4; vom Saal and Moyer, 1985). Thus, the 2F females potentially produce one or two fewer litters (about 20 mice) over their life span. However, this reproductive advantage of 0M is blunted, because their later litters have 50% more stillbirths, appar-

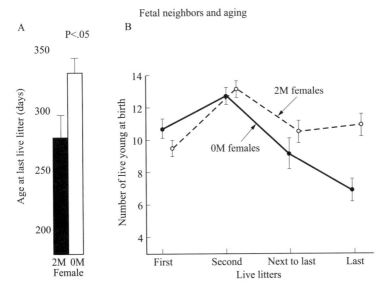

Figure 4.4. Female mice flanked by males (2M) undergo earlier reproductive senescence than 2F females. The mechanisms are not due to premature ovarian oocyte depletion, but rather involve increased stillbirths from delayed parturition in association with abnormal preparturitional plasma progesterone and estrogen levels (Holinka et al., 1979). Redrawn from vom Saal and Moyer (1985). A, Age at last litter is 75 days later in 0M females. B, 0M females have fewer live born pups in their later litters.

ently because of delayed birth (Fig. 4.4; vom Saal and Moyer, 1985). Some time ago, C.E.Fs laboratory showed that delayed birth in aging mice is associated with delayed onset of the preparturitional drop of progesterone, which facilitates uterine contractions (Holinka et al., 1979a,b). Additionally, we do not know if fetal neighbors modify the numbers of oocytes in the developing ovary or, by neuroendocrine programming, modify the rate of postnatal follicle atresia.

Age changes in reproductive behavior are also influenced by fetal neighbors, such that 0M females become behaviorally more like 2M females in response to injected testosterone at later ages (Rines and vom Saal, 1984). This observation is consistent with a model for cumulative continuing effects of endogenous sex steroids on neural loci during adult aging, such that the degree of exposure during prenatal development is additive with subsequent steroid exposure (Finch et al., 1980; Finch et al., 1984).

Male prostate characteristics are influenced by fetal neighbors, with implications that individual susceptibility to prostate diseases during aging may be traced to variations in fetal sex steroids from endogenous or external environmental sources (we emphasize that Fred vom Saal, not

C.E.F., a coauthor of vom Saal et al., 1994, is the originator of this hypothesis). Androgen receptor levels in 2F male fetuses are threefold higher than in 0F males (Even et al., 1992; vom Saal et al., 1997). Estrogen implants that increased fetal blood estradiol by 30% in 0M male fetuses caused severalfold *increases* in adult prostate cell numbers and androgen receptors per cell (Fig. 4.5A). Regional differences included more glandular epithelial buds in the urogenital sinus (vom Saal et al., 1997). Conversely, low doses of testosterone that did not alter fetal blood estradiol nonetheless *decreased* adult prostate size (vom Saal, Welshons, and Ganjam, unpub. obs., cited in vom Saal et al., 1994, p. 1237). The zone of the human prostate that develops benign

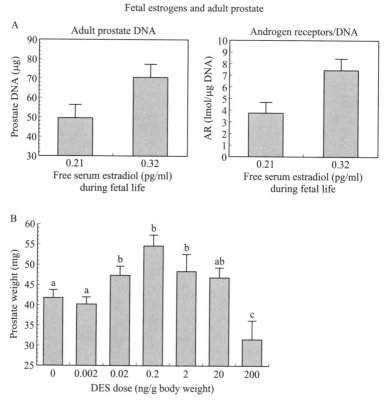

Figure 4.5. In mice, adult prostate characteristics are influenced by small differences in exposure to estrogens during fetal development. The data show an inverted U-shaped distribution Redrawn from vom Saal et al. (1997). *A*, Cell number (prostate total DNA content, *left*) and androgen receptor content per cell (receptor binding per μg DNA, *right*) are increased severalfold by implants of estradiol that approximate the difference between 0F and 2F males. *B*, Effects of diethylstilbestrol (DES) show a U-shaped dose–response curve on prostate.

prostate hyperplasia (BPH) and that may be sensitive to fetal variations in sex steroid levels is thought to be homologous to the rodent dorsolateral prostate (section 2.6.2).

Aging mice do not develop the same type of BPH found in humans (Maini et al., 1997). However, the prostate does enlarge in subgroups at the age of 18 months, which would be consistent with fetal neighbor effects. Although prostate tissue weight increased by 80% on the average, there was a multimodal distribution: one subgroup did not differ from 4-month-old males, whereas another subgroup had severalfold larger prostates (vom Saal, unpubl. obs., cited in vom Saal et al., 1994, p. 1238). Moreover, another genotype of mice at age 2.5 years had a subgroup with sexual performance (numbers of intromissions) that was similar to that of the young, whereas the remainder were impaired (Bronson and Desjardins, 1986). The impaired mice had disorders in various organs but no association with prostate enlargement. It has not been possible to follow fetal neighbor categories from birth to old age because of the high costs involved (granting agencies are often reluctant to fund experiments that take many years; maintenance of one mouse for its lifetime costs about $200). However, the proportion of aging mice retaining youthful characteristics approximated that of 2M males (15%). Because exposure of fetal mice to estrogen enhances male sexual performance as adults, as well enlarging the prostate, vom Saal speculates that 2M males should be resistant to these age changes.

How might slight variations in sex steroid levels have such profound effects on adult reproductive behavior, physiology, and aging? As mentioned in section 2.6.1, perinatal rodent brains are sensitive to major increases or decreases of sex steroid levels, which cause major effects on adult sexually dimorphic brain regions and on adult behaviors. Testosterone is converted to estradiol in the brain, where it regulates the proliferation of neurons, their survival, and their connections in the hypothalamus and other regions that control reproductive physiology and behavior. Although fetal neighbor effects on neuroanatomy are not reported in detail, there are indications of the expected effects of physiological variations in fetal sex steroid levels. The anterior hypothalamus, which contains sexually dimorphic nuclei, has 20% greater cytochrome oxidase activity in 2M versus 0M female gerbils (Jones et al., 1997), consistent with differences in neuron density (Fred vom Saal, pers. comm.). Hippocampal volume in gerbils also is associated with the fetal neighbor distribution (Sherry et al., 1996).

4.1.2. Humans

The fetal neighbor interactions may extend to humans, in the special circumstance of non-identical twins of the opposite sex. However, we emphasize that in humans, multiple births are the exception, not the rule (Table 4.1), which minimizes the impact of fetal neighbor effects

Table 4.1. Incidence of multiple births in humans and primates

Species	Major birth number (%)	Multiple births
Human	1 (>98%)[a]	MZ twins (0.4%)[b]
		DZ twins (1.2%)[c]
		triplets (0.013%)
		quadruplets (<0.00002%)
Great apes[d]		
bonobo (*Pan paniscus*)	1	rare
chimpanzee (*Pan troglodytes*)	1[e]	rare
orangutan (*Pongo pygmaeus*)	1	rare
gorilla	1	rare
Lesser apes		
Hyalobates (gibbons), 9 species	1	rare
Other primates (examples)		
marmosets (*Callithrax*)[f]	2–3	normative
mouse lemurs (*Microcebus*)	2–3	normative

[a]Slight differences among human populations (98.5–99.3%), due to variable DZ twinning (Hrubec and Robinette, 1984; Bulmer, 1970).

[b]MZ incidence is relatively constant across populations and independent of maternal age.

[c]DZ twins are more common in Africans (1.5%) than in Asians (0.7%; California 1905–1959). The incidence of DZ twins increases fivefold with maternal age (Hrubec and Robinette, 1984) and may become even more common due to in vitro fertilization.

[d]The general description that twins or higher multiple births are "rare" in the great apes (Nowak, 1994) implies that >95% of births are singletons.

[e]Chimpanzee births are predominantly singletons, although occasional pedigrees show heritable twinning (Geismann, 1990).

[f]Poole and Evens (1982) summarized data from four colonies of the common marmoset (*Callithrax jacchus jacchus*). Triplets are the common birth frequency (55% triplets versus 35% twins versus 2% singletons). Maternal neglect usually eliminates one neonate. Triplets rarely survive unless assisted by humans.

on the populations. Like our great ape relatives, human births are almost 99% singletons from one egg issued from either ovary and fertilized in the single uterine chamber. Less commonly, 0.4% of births, are identical twins arising from the same egg that split after fertilization to yield identical, monozygous (MZ) co-twins. If two eggs are fertilized separately, the resulting dizygous (DZ) twins can be of the same sex or opposite sex, depending on the sperm.

In contrast to DZ twins, which always have separate placentas and chorionic membranes, MZ twining can occur at different times in early development, which leads to chance variations in sharing or not of chorionic membranes and access to maternal blood. In about one-third of MZ twins, the fertilized egg divides early, yielding separate fetal membranes. However, the majority of MZ twins form later, so that the embryos share an outer chorionic membrane, within which each fetus is in a separate amniotic sac. Monochorionic MZ twins are smaller at birth by 4–9% than dichorionic MZ or DZ twins (Phillips, 1993) and also more variable in growth and birth weight, possibly because of competition for the maternal blood supply (Phillips, 1993). The fetal skin is permeable to some solutes up through the 18th week of pregnancy, which may allow exposure of the fetus to co-twin androgens by diffusion. There are no data on fetal sex steroid levels in human twins. The following studies did not distinguish whether the MZ twins were mono- versus dichorionic twins.

Women of opposite sex twins do not differ markedly in fecundity from women of same sex twins (Lochlin and Martin, 1998). This study of women had a very large sample size (600 opposite-sex DZ pairs, 1400 MZ pairs) and was based on responses to a mailed questionnaire. For example, the length of menstrual cycles, age at menarche, age at first pregnancy, and number of full-term pregnancies differed by $< 4\%$ mean values. However, opposite-sex co-twins reported modest trends for relatively more premenstrual sleep disturbances (20%) and premature births (30%) relative to mean values for same-sex co-twins. The apparent absence of major effects on human co-twin reproduction does not rule out subtle effects that may be obscured in open human societies. Recall the influence of social environment on the cycle length in mice, which include conditions that suppress fetal neighbor effects (Fig. 4.3B). Fetal neighbor effects on the age of first (pubertal) mating are also sensitive to the social environment (vom Saal and Bronson, 1978). Moreover, the self-reports of menstrual cycle lengths cannot be considered conclusive, unless based on longitudinal records (Treloar et al., 1967).

Last, normal co-twins of the opposite gender did not, as children, differ in sex-typed play (Henderson and Berenbaum, 1997), using criteria that showed the effects of the masculinizing adrenal hyperplasia (CAH) syndromes (Fig. 2.11). In CAH, elevations of androgens from genetic defects in adrenal steroid synthesis or from maternal adrenal tumors reach the male range in female fetuses. The resulting masculinized external genitalia are usually surgically corrected. CAH girls consistently have less gender stereotyped play and other gender-sensitive activities than sisters in the same family (Berenbaum, 1999).

Nonetheless, there is evidence for co-twin interactions that influence the development of the inner ear and lead to a curious type of sound production by the ear itself. In about 35% of normal individuals, the huma cochlea emits weak tones of < 20 decibels, the "spontaneous otoacustic emissions" (OAEs), which are continuously produced and propagate retrogradely through the middle ear into the external canal. At their low amplitude, spontaneous OAEs are inaudible and do not interfere with hearing, but rather are associated with hearing sensitivity (McFadden and Mishra, 1993). Each individual has a distinctive OAE phenotype from birth onward (McFadden, 1993). Spontaneous OAEs in the 0.5–6 kHz range are two- to threefold more common in females than males (Probst et al., 1991; McFadden, 1993). The vertebrate cochlea produces other sounds besides spontaneous OAEs, some of which may be evoked by external sounds, for example, the click-induced OAEs that are evoked in all humans with normal cochleas and middle-ear systems (Probst et al., 1991; McFadden and Pasanen, 1998).

Females with a male co-twin have masculinized patterns of OAEs, whether spontaneous or click-induced. Young adult female co-twins had 50% fewer OAEs than if the DZ co-twin was female, and about the same number as males in general (Fig. 4.6A). These interactions may be mediated by fetal testosterone, which is transiently severalfold higher in males than females by 9–16 weeks (Fig. 2.12). The neural crest cells, from which the cochlea develops, arise by 3–4 weeks. The cochlea is partially developed during the time of higher fetal male testosterone but does not acquire the sound-detecting hair cells until after week 21. The sex difference in OAEs may be caused by direct effects of testosterone on the cochlea, or indirectly through effects of testosterone on neural input to the cochlea from the brainstem (olivocochlear afferents) that can modify OAEs (McFadden, 1993; Probst et al., 1991). Direct effects of estrogens on OAEs of the adult brain are indicated by a case report of one transsexual male (McFadden et al., 1998).

Sexual orientation has been associated with the size of click-induced OAE in females but not males (McFadden and Pasanen, 1998). In volunteer respondents who were not from a twin sample, the self-reported heterosexual females had click-induced OAEs with 50% higher amplitude than in males, consistent with prior data (Fig. 4.6B). Those women with self-reported homosexual orientation had signals with 25% lower average amplitude, whereas homosexual men did not show this effect. The small sample sizes ($N = 11–57$) of course prevent strong conclusions. Drug use and lifestyle may be factors; for example, OAEs are diminished by high doses of aspirin and exposure to intense sounds (McFadden and Pasanen, 1998).

Whether hormonal effects of normal or abnormal conditions of pregnancy influence adult sexual behavior is highly controversial. By the standards of the fetal neighbor interactions in rodents, the high androgen levels experienced by CAH girls have a quite subtle impact on postnatal behavior. However, it would be rash to conclude from these inconsistent findings that the human brain is not sensitive to variations in fetal sex steroids. As McFadden and Pasanen (1998) point out, the 40–70% estimated heritability of homosexual orientation in identical twins allows for fetal interactions and other less tangible environmental effects. Chance variations in fetal sex steroid levels (Fig. 2.12), for example, could vary the penetrance of putative genes for sexual orientation by influencing the development (numbers of neurons or synapses) in sexually dimorphic neural circuits (Fig. 1.8). The powerful pheromonal effects on manifestations of fetal neighbor interactions in rodents (Fig. 4.3) also show the need for exacting definition of environmental factors in studies of humans. Despite the well-documented sexually dimorphic brain regions and behaviors, investigators face great difficulties in identifying *which* of

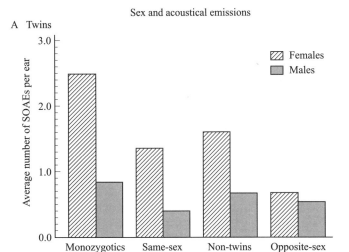

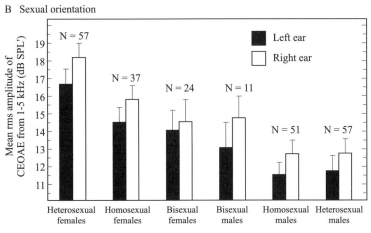

Figure 4.6. Gender differences in the incidence of spontaneous otoacoustic emissions (SOAEs) which are two to three times more common in female humans than males at all ages. *A*, The incidence of spontaneous OAEs (SOAEs) approximates the lower values of males in DZ twins of the opposite gender, implying transplacental influences of sex steroids. Redrawn from McFadden (1993). *B*, The amplitude of click-induced OAEs (CEOAEs) is largest in heterosexual women and smallest in heterosexual men. This graph shows CEOAEs induced by 75 dB clicks; other click intensities gave similar profiles. Women who are homosexual by self-report have smaller amplitudes, which implies a partial masculinization of this trait in utero. The larger amplitudes of CEOAEs in bisexual men than in homosexual men do not fit a simple hypothesis; the small sample sizes (given above bars) make conclusions provisional. Redrawn from McFadden and Pasanen (1998).

the many loci in these complex circuits are pertinent. Functional brain imaging may give important information.

In conclusion, there are possible effects of co-twins of the opposite sex on each other's neuronal development in OAEs. However, these findings do not imply that a person's sexual orientation is developmentally caused by a pattern of OAEs, or that OAEs could be used to predict adult sexual orientation or behavior. The significance of gender and co-twin differences in OAEs is difficult to judge, because only a subgroup of adults demonstrate spontaneous OAE phenomena. Functional brain imaging (fMRI, etc.), which is being used to study sex differences and effects of estrogen replacement in brain functions during cognitive tasks, might detect co-twin effects on adult brain functions in sexually dimorphic regions and circuits. Further studies on marmosets (Abbott, 1984) and other laboratory primates with multiple births are also needed.

4.1.3. Evolutionary questions

The evolution of reproduction in mammals that regularly have litters with two or more adjacent fetuses must have led to mechanisms to prevent impaired reproductive organ development by sex steroidal cross-talk between resident fetuses of opposite sex. This issue does not arise in armadillo species, which have MZ quadruplets (all same sex; Hardy, 1995; section 1.5.2). Rodents and other species with multizygous pregnancies have evolved separate placentas with restricted interfetal exchange through direct circulation or diffusion. These adaptations help to maintain the distinctly different steroid levels of male and female fetuses required for normal reproductive development (vom Saal, 1984).

Mechanisms of interfetal isolation are also important to other species that more rarely have DZ twins of the opposite sex, such as humans and great apes. However, these mechanisms are not found in all other species with occasional twinning. For example, in domestic cattle, which generally have one calf, the occasional DZ female twin of an opposite sexed pair is usually sterile. These "freemartins" have abnormal reproductive tracts, often with stunted ovaries and seminiferous tubule-like entities containing Sertoli cells (Marcum, 1974; Behringer, 1995; Jankowski and Ilstad, 1997). Freemartins are not caused by steroidal cross-talk between opposite-sex fetuses. Rather, their abnormal development is due to cellular chimerism from germ and marrow cells transferred from the other embryo via vascular anastomoses in the chorioallentoic membranes. These variations among opposite-sex twins may be due to chance variations in the development (angiogenesis) of cross-circulation, which, in turn would allow different degrees of cell traffic and chimerism.

There are major differences in uterine vascular anatomy among species that have multizygous births, which suggest that maladaptations from fetal neighbor effects lead to the evolution of different mechanisms that allowed multiple births. It is well documented that the numbers of offspring can change rapidly in response to selection for different mortality patterns of neonates, juveniles, or adults (Rose, 1991; Austad, 1993). The singleton birth mode of humans, the high dependency of neonates on maternal care for many years, the late puberty, and the multidecade life span potentials are also characteristics of the extant apes (Finch and Sapolsky, 1999). These extended life histories appear to have selected against multizygous births, which are a major drain on maternal resources (Table 4.1, note f). In view of this apparent several million years of singleton births in human evolution, it seems remarkable that the capacity for viable twin development has persisted, because twinning is relatively minor in modern humans. Also, recall from above that neonatal twins are smaller and have higher mortality. From this perspective, the persistence of twinning in human evolution is puzzling (recall from above that DZ opposite-sex twins have normal fecundity). Because unused traits are often rapidly lost during evolution (Diamond, 1986; Finch, 1990, p. 639), the persistence of low-level twinning in humans implies pleiotropic interactions with other life history trait genes. DZ twinning rates vary among human populations, and are relatively high in West Africa which implies a selective advantage (Table 4.1, note c), in contrast to MZ twinning, which is relatively constant.

How might the fetal neighbor effects have become adaptive in rodent microevolution? vom Saal (1983) proposed that the variations of aggression and other behaviors from fetal neighbor effects could be adaptive, because they yield a diversity of phenotypes from the same litter. Thus, depending on local population density, the degree of aggressiveness and the age at first mating could favor reproductive success in a microevolutionary context. These features were incorporated into a model based on chaos theory, which showed that population size can oscillate about a stable value for population density (Cowell et al., 1998). With the parameters chosen, the periodicity of the oscillations approximates the several-year intervals between population crashes observed in natural populations of voles and other microtene rodents. As Real (1980, p. 47) has observed:

> When fitness is uncertain, organisms will engage in diverse behaviors whose covariances are negative; while under "fitness certainty," the most adaptive strategy will be a single type of behavior.

More generally, we argue that the chance variations in brain structure (section 1.4) will prove to cause variations of information processing as well as of neuroendocrine functions. Nongenetic variations can be adaptive in two regards. First, like the fetal neighbor interactions, they provide a reservoir of individual differences in populations that are independent of selection or the degree of inbreeding. These reproductive strategies of microtene rodents are very different from those that evolved in long-lived primates, which have much lower population density and much greater parental investment.

4.2. Environmental estrogens

Estrogens from the external environnment can also profoundly influence the development of the reproductive system in mammals. The adverse consequences of the drug DES (diethylstilbestrol) on increased cancer risk are a precedent for considering the effects of many other hormone-like substances in the environment. Many xenoestrogens (estrogen-like substances, both natural and industrial) to which individuals may be exposed by chance, act as "endocrine disruptors" at very low levels (vom Saal et al., 1998). Organochlorine compounds are of particular interest because of their estrogen-like effects (e.g., Wade et al., 1997). Recently, blood levels of the pesticide dieldrin, a weak estrogen, were associated with twofold higher risk of breast cancer by a 17-year follow-up in the Copenhagen City Heart Study (Hoyer et al., 1998). vom Saal et al. (1998) suggest that there may be no safe threshold for levels without permanent (organizing) effects on developing tissues. In contrast, most evaluations of toxins have assumed some threshold level, below which traces are inconsequential and therefore "safe." An emerging issue concerns to what degree environmental steroids from diet and industrial sources cause population differences in prostate cancer, which is the main cancer in men. For example, in Los Angeles, the apparent incidence (diagnosis rate) of prostate cancer is 40% higher in Blacks than non-Hispanic whites (Danley et al., 1995). It is unclear whether these differences represent genetic susceptibility or exposure to carcinogens.

Diethylstilbestrol (DES) is a major example of a drug that crosses the placenta to cause lasting effects that interact adversely with age-related changes in the reproductive system. For about 30 years after its approval by the U.S. Food and Drug Administration in 1941, DES was widely prescribed to pregnant women to reduce risk of miscarriage, premature labor, and diabetic complications (Herbst and Bern, 1981; Newbold and McLachlan, 1996). Moreover, starting in 1954, DES was widely fed to cattle and poultry to enhance growth. In 1971, DES exposure during pregnancy was first associated with a high risk of a rare malignancy in DES-exposed daughters, the clear cell vaginal adenocarcinoma (Herbst et al., 1971). Nonetheless, DES was finally banned for the U.S. domestic meat industry only in 1979.

DES, either fed or injected into pregnant rodents, readily crosses the placenta to cause potent effects on the development of the male and female reproductive tracts. Table 4.2 summarizes effects of DES on mice that model major features of human DES pathology. The cells giving rise to vaginal adenocarcinoma may be derivatives of the müllerian duct, from which develop the uterus, cervix, and cranial portions of the vagina (Fig. 2.11).

By middle age (12–18 months), DES-exposed fetuses had 50-fold more reproductive tract abnormalities, including vaginal tumors and cysts in wolffian ducts (Haney et al., 1986; McLachlan et al., 1975; Iguchi et al.,

Table 4.2. Adult lesions caused by prenatal exposure to diethylstilbestrol (DES)

Lesions	Mice	Humans
Females		
infertility (many causes)	Newbold and McLachlan, 1996	Newbold and McLachlan, 1996
malformation of oviduct	Newbold et al., 1983a	DeCherney et al., 1981
premature loss of oocytes	Newbold et al., 1983a	
neuron numbers modified during development	Döhler et al., 1994; Faber et al., 1993	
ovarian cysts	Newbold et al., 1983a	Haney et al., 1986
vaginal adenocarcinomas and benign tumors	Newbold and McLachlan, 1982	Herbst et al., 1971; Hanselaar et al., 1997
Males		
infertility (diverse causes)	Newbold and McLachlan, 1996	Newbold and McLachlan, 1996
undescended testes (cryptorchidism)	McLachlan et al., 1975	Bibbo et al., 1977
müllerian duct lesions	epididymal cysts; nodules in area of prostatic utricle McLachlan et al., 1975; Newbold et al., 1987	enlarged prostatic utricle; metaplasia in prostatic ducts; no evidence for interactions with BPH Driscoll and Taylor, 1980
testicular malignancy	5% of DES-exposed mice Newbold et al., 1985	Gill et al., 1976, 1979

Adapted from Tables 3 and 4 of Newbold and McLachlan (1996).

1995). We suggest that the occasional reproductive tract abnormalities in control groups of mice used in studies of exogenous estrogen (vaginal tumors, ovarian cysts) represent endogenous variations in sex steroid levels or target cell responses that reach a threshold to overlap with the DES effects. DES may also modify ovarian aging, because DES-exposed mice have premature depletion of ovarian oocytes by 6 months (Newbold et al., 1983a,b). There are no data on reproductive aging in DES-exposed human mothers or their daughters. Relatively subtle initial damage to the primary follicular pool may take years to emerge, because of the remarkable compensatory ability of follicle maturation (Hirshfield, 1997).

Males are also vulnerable to DES. Sterility and cryptorchidism (including undescended testicles) are common. Human infants exposed to DES during pregnancy have more frequent occurrence of prostatic utricle and of dilated prostatic ducts filled with metaplastic (squamous) cells (Driscoll and Taylor, 1980). Fetal mice exposed to DES have a high incidence of abnormal growths in the adult prostate (McLachlan et al., 1975; Newbold et al., 1987). DES has lasting effects on prostate size, with an inverted U-shaped dose–response curve (Fig. 4.5B; vom Saal et al., 1997). This complex dose–response curve is in part due to inhibition by DES of the normal regression of the müllerian duct system, as well as to stimulation of cell proliferation. Although the müllerian duct remnant contributes to a prostate subregion with BPH during aging (section 2.6.2), the DES-induced nodules may not be equivalent to those of BPH. In humans, no interactions of DES with BPH or prostate cancer are recognized (Newbold and McLachlan, 1996). No consistent effects of DES exposure on cognition or behavior have been found in humans (Berenbaum et al., 1995; Lish et al., 1992; Pillard et al., 1993). Such effects were anticipated from the masculinizing effect of DES on hypothalamic neuronal numbers in rats treated prenatally (Döhler et al., 1984; Faber et al., 1993).

The postnatal replication of Sertoli cells in the testis is influenced by testosterone, but also by thyroid and gonadotropic hormones. Environmental toxicologists are concerned with many steroid-like agents that could alter spermatogenesis. The numbers of Sertoli cells surrounding each spermatogenic gonocyte are considered determined by puberty, when Sertoli cells become postmitotic. Because each Sertoli cell supports a finite number of sperm during gametogenesis, the number of Sertoli cells per testis is thought to limit the maximum potential sperm production, and thus may prove to be a lifetime endowment, like that of the ovarian oocytes.

Industrial estrogenic compounds at low, environmentally relevant doses (parts per billion) permanently alter development in rodents. For example, the bisphenol used in plastic dental sealants can leak into the digestive tract and circulation. When fed to pregnant mice, bisphenol at the trace dose of $2 \mu g/kg$ caused 35% increases in adult prostate weight (Nagel et al., 1997). Effects of bisphenol and other nonclassical estrogens on prostate aging are not reported.

These examples of estrogenic drugs and industrial chemicals are precedents for global effects of chance interactions with environmental endocrine disruptors that modify the development of the reproductive system. It is likely that many of these agents will show activity even at the lowest doses that are within physiological variations, as shown for effects of estradiol on prostate growth (vom Saal et al., 1997). Epidemiological studies of prostate cancer and other reproductive tract disorders of aging that differ among human populations are likely to implicate many other natural and industrial chemicals. But full understanding of these population variations will require information on how external agents interact with the variations in cell number during development that arise from chance processes that are largely endogenous.

4.3. Stress and effects on the nervous system

Chance exposure to stress during development can also alter adult functions and outcomes of aging just as fetal neighbor effects and fluctuating asymmetry can. Many stressors activate the hypothalamic-pituitary-adrenal axis (HPA axis) and cause sustained elevations of blood glucocorticoids (Sapolsky, 1996). Nonetheless, stressors act through a great range of mechanisms that involve the HPA axis to varying degrees.

4.3.1. During pregnancy

Exposure to constant bright light for successive days is a strong stressor for nocturnal rodents that has debilitating effects. In particular, light-induced elevations of corticosterone during pregnancy converted 2F females toward the 2M characteristics of more aggressive behavior and longer estrous cycles (vom Saal et al., 1990). In this situation, stress may reduce phenotypic diversity, which may be adaptive at certain population densities (see section 4.2; vom Saal et al., 1998a). Moreover, offspring of stressed pregnant rats have impaired memory and altered hippocampal neuron functions (Day et al., 1998). The mechanisms are not known, but may be indirect, for example, glucocorticoids, which are elevated by many stressors, can inhibit glucose uptake in neurons (McEwen and Sapolsky, 1995). Stress also modifies maternal behavior, as described below.

Fluctuating asymmetry (FA), the individual variations in the size of bilateral anatomic features (section 1.5.3), is increased by stress during development in many species (Parsons, 1992; Møller and Swaddle, 1997). These effects may have unpredicted consequences for prey–predator relations in food chains and other species interactions. The following examples represent outcomes of chance fluctuations in the environment. Returning to bristle number in fruitflies (Table 1.4), bristle FA was increased twofold by exposure during development to sublethal temperatures (30° vs. 25°C); offspring from older mothers also had increased FA (Parsons, 1990). The pesticide dieldrin (a weak estrogen; section 4.2) also increased bristle FA in an insect that infests sheep, the Australian blowfly *Lucilia cuprina* (McKenzie and Yen, 1995). FA is among other sublethal effects of pesticides on insect development that increase vulnerability to predation.

In some birds the symmetry of elongated tail features, a sexual dimorphism and a mate attractant, indicates individual fitness. For example, in barn swallows (*Hirundo rustica*), the amount of rainfall during the molting season influences FA, such that females prefer males with the most symmetric features (Møller, 1996a; Swaddle and Witter, 1994). To test the relationship of chance in food availability to FA, female European starlings (*Sternus vulgaris*) were subjected to food restriction under conditions of molting, when new tail features are grown (Swaddle and Witter, 1994). The experiment lasted 15 weeks, during which birds were either fed ad libitum or deprived of food each day, either regularly in the morning or evening, or unpredictably on one of these times from day to day. Food deprivation, when it was unpredictable, caused twofold more FA. Moreover, an index of ovarian follicular maturation (spottiness of chest feathers) and the size of subcutaneous fat depots were associated with lesser FA. This experiment indicates that unpredictable food supplies are more stressful than transient predictable food deficits and that, through unknown neuroendocrine mechanisms, influence ovarian development and the growth of feathers.

In mammals, various stressors during pregnancy increase the bilateral asymmetry of teeth (dentitional FA). This may occur in offspring of pregnant rats exposed to heat stress (Siegel et al., 1997), in offspring of diabetic monkeys (Kohn and Bennett, 1986), and in offspring of alcoholic women (Kieser, 1992). Although these stressors have different pathophysiologies, there could be final common pathways, from which the organ locations of FA might be anticipated. For example, individuals with cleft palate had greater dental FA (Adams and Niswander, 1967), which is consistent with the neural crest origin of teeth and many other craniofacial components. Returning to an earlier example, it would be interesting to examine dental FA in twins with hippocampal volume asymmetry (Fig. 1.6). As discussed in section 1.5.3, an IQ measure in college students showed a partial, negative correlation with FA (Furlow et al., 1997). These observations defy simple interpretation, because of deleterious social-psychological consequences that may affect a child whose appearance departs even slightly from the norm. In adults, facial attractiveness is correlated with slight facial asymmetry (Grammar and Thornhill, 1994; Thornhill and Gangestad, 1993), while more severe facial and breast asymmetry are negatively correlated with attractiveness and reproductive success (Møller et al., 1995; Thornhill et al., 1995). Putting these examples together, we suggest that chance exposure to stressors during pregnancy could influence outcomes of aging by increasing FA in adult brain regions that lower the cellular reserve. Together with the examples of stress effects on fetal neighbors in rodents, these few examples hint at the deep engagement of phenotypes with environmental factors that are superimposed upon chance processes during development.

4.3.2. Neonatal handling and later neuron loss

We briefly expand our focus on prenatal effects to describe a perinatal effect on adult stress responses that plays out in hippocampal aging. Adult rat responses to stress are conditioned by neonatal handling, with lasting effects on adult brain regions that control stress responses. If gently handled by humans for 15 minutes each day as neonates, young adults are less reactive to stress, as measured by smaller HPA activation and faster return to baseline levels of glucocorticoids (Liu et al., 1997; Sapolsky, 1997; Anisman et al., 1998). As adults, the handled rats showed striking inverse correlations between the amount of maternal grooming and the individual neuroendocrine responses to restraint stress: elevations of adrenocorticotropin (ACTH) and glucocorticoids (corticosterone, 'B') (Fig. 4.7A). In unstressed rats, the amount of maternal grooming also

Maternal grooming and stress response

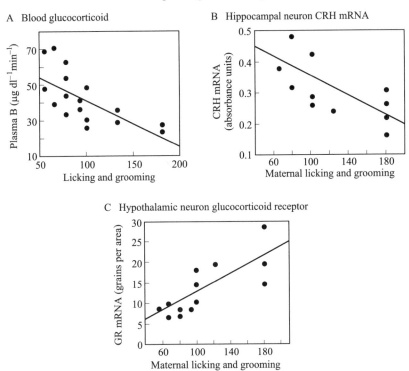

Figure 4.7 Maternal grooming of neonatal rats attenuates their stress-induced hypothalamic-pituitary adrenal axis activation by restraint stress as adults. Redrawn from Liu et al. (1997). *A*, Restraint stress-induced elevations of blood and corticosterone ("B") are smaller in adult rats that received more licking and grooming as neonates. The integrated exposure to corticosterone (A_3) varied inversely with the amount of licking and grooming received as neonates. *B*, Neuronal content of corticotropin-releasing hormone (CRH) mRNA in the paraventricular nucleus of the hypothalamus that regulates pituitary secretions of ACTH; CRH mRNA varied inversely with licking and grooming received as neonates. *C*, Neuronal content of glucocorticoid receptor (GR) mRNA in the hippocampus, a target of corticosteroid-induced neuron damage during aging.

correlated with the corticotropin releasing hormone (CRH) mRNA per neuron in the hippocampus (Fig. 4.7B). CRA is the hypothalamic neuropeptide that controls the release of adrenocorticotropin (ACTH) from the pituitary. In addition to its role in declarative and spatial memory, the hippocampus also influences the output of CRH through feedback to the HPA axis. The smaller HPA activation and greater sensitivity to feedback in shutting off the HPA axis after stress thus involve a resetting of gene activity in neurons of the hippocampus and hypothalamus. Levels of glucocorticoid receptor (GR) mRNA in hypothalamic neurons were higher in proportion to maternal grooming

Mechanisms in stress responses

A Variations in feedback

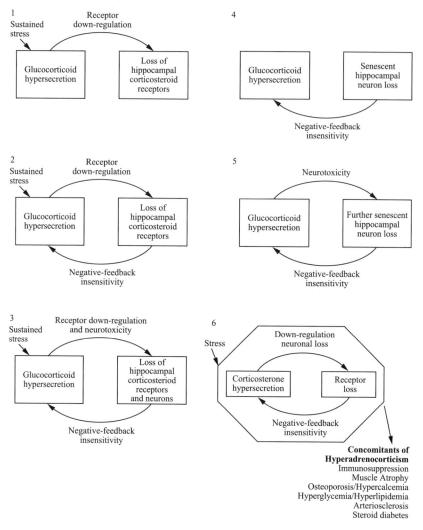

**Concomitants of
Hyperadrenocorticism**
Immunosuppression
Muscle Atrophy
Osteoporosis/Hypercalcemia
Hyperglycemia/Hyperlipidemia
Arteriosclerosis
Steroid diabetes

B Effects of neonatal handling

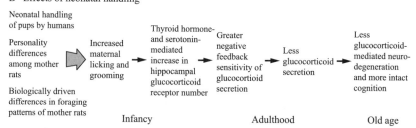

(Fig. 4.7C). In Sapolsky's (1997) scheme for these complex interactions (Fig. 4.8), a cascade of programming neural circuits may be adaptive in the short-term, but is costly in outcomes of brain aging.

The mechanisms through which the effects of handling are mediated remain unknown. Mothers became more attentive to the handled pups, doubling their grooming efforts (Liu et al., 1997). The handling of pups stimulates ultrasonic vocalizations, which may stimulate maternal attention (Lee and Williams, 1974). Daily 15 minute handling of neonatal rats by humans may not be a deprivation because nursing rats routinely leave their nests for similar times; the removal of pups from the natural environment of a dark burrow, however, is "unsettling" (Liu et al., 1997). Females also differ in their response to apparently endangered pups, which Sapolsky (1997) described as "personality differences between mother rats" (Fig. 4.8B). Fetal neighbor effects could impinge on individual maternal differences in response to their pups and in pup responses to maternal attention, but were not included in these studies. Indications of such effects are more aggressive behavior by 2M than 2F females in response to a strange female intruder, and more aggressive grooming and tail rattling during direct female–female competition, as described in section 4.2.1.

The handling of infant rats interacts with the vulnerability to hippocampal damage during aging. The briefer HPA activation and elevations of glucocorticoids in response to stress as young adults imply reduced lifetime exposure to glucocorticoids. Considerable evidence shows that hippocampal neurons of humans can be damaged and even die as the result of sustained stress (Sapolsky, 1996; McEwen, 1997). There is provisional evidence that the size of the human hippocampus is inversely correlated with the duration of combat stress and with the elevations of glucocorticoids in Cushing's syndrome (Sapolsky, 1996). The smaller hippocampus implies atrophy due to shrinkage of individual neurons, as well as neuron loss. Both changes are observed in chronic exposure to high levels of glucocorticoids (Sapolsky et al., 1990). Moreover, the handling of rats, which blunted the HPA responses of young adult rats to stress (Fig. 4.7), was also associated with less hippocampal neuron loss in old age (Meany et al., 1988; Fig. 4.9). These findings complement observations that astrocytic hyperactivity in aging rodents is correlated with their degree of blood corticosteroids, and that hippocampal neuron loss and astrocytic hyperactivity can be reduced in old rats that are adrenal-

Figure 4.8. *A*, The hypothalamic-pituitary-adrenal (HPA) axis, showing different feedback settings and interactions with the hippocampus. *B*, A schema for maternal–neonatal interactions that program adult responses to stress and eventually the vulnerability to corticosteroid-dependent neuron damage during aging. Redrawn from Sapolsky (1997).

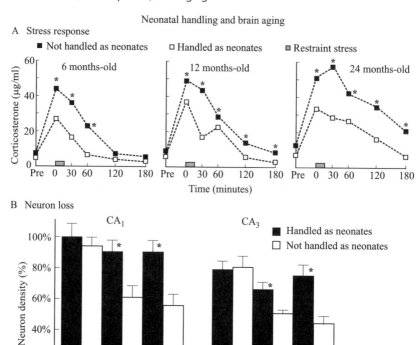

Figure 4.9. Neonatally handled rats show briefer elevations of corticosterone in response to restraint stress as young adults, *A*. The implied greater lifetime exposure to corticosterone was associated with greater hippocampal neuron loss during aging, *B*. Redrawn from Meaney et al. (1988).

ectomized and allowed to age with low exposure to glucocorticoids (Finch and Landfield, 1985; Landfield, 1994).

Whether the effects on neonatal rodents pertain to outcomes of aging in humans is, of course, speculative. There is little longitudinal data on how early parental attention influences adult reactions to stress. Even so, aging humans differ in their glucocorticoid levels. Among healthy older people living in communities (the MacArthur Foundation Network on Successful Aging), a subgroup was defined by its progressive elevations in glucocorticoids. This subgroup also had a greater incidence of cognitive decline, which implies impaired HPA feedback (Seeman et al., 1997).

In conclusion, it is plausible that early adult patterns of stress response have a later impact on brain functions. In nature, these costs to rodent brain functions would not be incurred because of the generally short life expectancy (section 1.7.3). Humans and other long-lived animals may have evolved mechanisms to protect their neurons against many kinds of stressors, at least during their major phase of reproduction when natural selection acts most strongly. The chance variations of maternal attention are another example of chance outcomes of development that influence adult functions and aging.

4.4. Nutrition during pregnancy

By bad luck, individuals may develop in malnourished mothers during famines caused by crop failures, embargoes, or war. The adverse effects of maternal malnutrition on adult health in studies on rodents can extend over several generations. However, the findings on humans are controversial and may not be reconciled without information on elusive parameters, such as the composition of transition diets during restoration of normalcy, pathogens, and psychosocial factors, which can have potent effects on the outcomes of infections and chronic diseases. We chose a few examples from this important but controversial literature.

Effects of malnutrition on adult functions can extend from a single episode of malnutrition across generations (transgenerational effects). For example, the offspring of malnourished pregnant rats had higher blood pressure as adults (Langley and Jackson, 1994). The experiment adapted female rats to a threefold range of protein intake to maintain normal litter size, but after birth placed nursing mothers on the same breeder chow diet (20% casein). Despite access to optimum protein thereafter, the adult systolic blood pressure remained perturbed and varied inversely with maternal protein deficiency (Fig. 4.10). The body size differences at birth also persisted as adults. In another study, female offspring of diabetic rats had impaired glucose tolerance as adults; when mated, their offspring (third generation) still showed hyperglycemia (van Assche and Aerts, 1985). In this example, transgenerational effects were attributed to persistent diabetes from one generation to the next, by which the fetal pancreas becomes hyperactive. An extreme experiment with 12 consecutive generations of malnourishment showed remarkable organ specificity: the pancreas and kidneys were about 40% smaller, whereas the size of brain and eyes are almost normal. The malnourished lineage was hyperexcitable and less effective in problem solving (Stewart et al., 1975). As might be anticipated, protein deficiencies during pregnancy in rats followed by a normal diet after weaning also shorten the life span of adults, in association with poorer glucose tolerance (Desai et al., 1995; Hales, 1997).

In the studies described above, the pups were also nursed by their malnourished mothers. To resolve post- and prenatal maternal and fetal effects of malnutrition, rats born from malnourished dams were cross-fostered on ad libitum fed nurses (Zamenhoff et al., 1971). In the second generation of females, brain cell number (DNA content of cerebral hemispheres) was 7% below normally fed controls. These effects

Rat fetal malnutrition and adult blood pressure

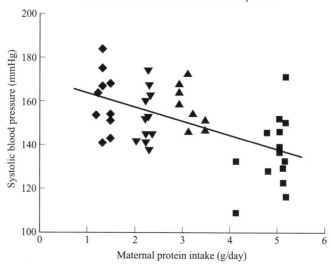

Figure 4.10. Malnutrition during pregnancy in rats increases blood pressure in offspring as adults. Female rats were adapted to a threefold range of protein intake (6–18% casein) that did not alter the litter size. The 6% casein had the greatest impact on body size at birth (−30%). The nursing mothers were placed on the same breeder chow diet (20% casein). Despite access to optimum protein, size differences persisted as adults. The scatter plot shows the four different levels of maternal malnutrition (four different symbols), in which the ranges of individual values overlap. Redrawn from Langley and Jackson (1994).

were transmitted maternally, because offspring of cross-fostered males from the same experimental group of malnourished grandmothers were normal. Besides a possible role of uterine vascularity raised by more recent studies, Zamenhoff et al. (1971) suggested that the pups had deficient nursing behavior. These and other studies show that malnutrition during pregnancy can have multigenerational effects that *may not* be rectified by improved nutrition within a single generation.

Although these reports describe adult deficiencies resulting from malnutrition during development, there are no data on life span. In fact, caloric restriction of adult rats with adequate micronutrients has positive effects that *slow* many aging processes. More than 100 studies document that postpubertal rats respond to 20–40% reduction of ad libitum intake of food which increased the life spans, in assocation with slower mortality rate accelerations (Finch, 1990, p. 508) and delay of age-related diseases (Finch, 1990; Sohal and Weindruch, 1996; Weindruch and Walford, 1998; Masoro and Austad, 1996; Finch and Morgan, 1997). Thus, caloric deficits can have very different effects depending on when they occur during life history.

In recent findings on humans, many populations show an association of low birth weight and greater incidence of coronary artery disease (CAD), stroke, and diabetes (Barker, 1994, 1995; Forsén et al., 1997; Rich-Edwards et al., 1997; Scrimshaw, 1997; reviewed by Waterland and Garza, 1999). These populations include those in England, Sweden, India, and the United States (Nurses Health Study). A review of 34 studies showed an overall consistency in the inverse correlations of low birth weight with systolic blood pressure of adults and neonates (Law and Shiell, 1996). Nonetheless, in the Nurses Health Study, low birth weight was associated with <2% of the nonfatal CAD (Rich-Edwards et al., 1997; Fig. 4.11). It is possible that the association of birthweight with vascular disorders is "driven by small proportions (<12%) of women at extremes of the birthweight range," (Rich-Edwards et al., 1997) that is, in babies weighing <5 lb and >10 lb (Fig. 4.11). The majority in the middle range of 5–10 lb birth weight showed no influence of birth weight on adult risks of CAD or stroke. In chapter 5, we discuss birth-related stress and cardiovascular lesions.

Human birth weight is highly sensitive to maternal nutrition, even within the normal range of ad libitum intake. Complex associations of voluntary food intake were found in a prospective study of white women from Southampton (Godfrey et al., 1996). Carbohydrate (CHO) intake in early pregnancy was *inversely* associated with birth weight and placental weight; for example, for CHO intake in excess of 340 g/day, the birth

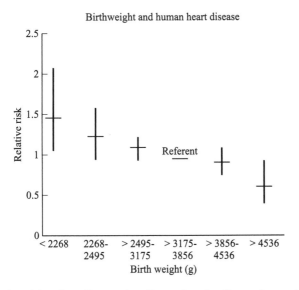

Figure 4.11. Risks of cardiovascular disease (nonfatal) vary inversely with birth weight. Data from Nurses Health Study (U.S.): 70,297 women, aged 30–55; mean age, 42.1 years. Redrawn from Rich-Edwards et al. (1997).

weight was 120 g (3.5%) lower than that of the lowest CHO intake group (< 265 g/day). However, there were no consistent statistical associations among placental size, fetal growth, and birth weight (McCrabb et al., 1991). A caveat is that human twins, which have smaller birth weight by 900 g (−20%), do not differ in life expectancy from the general population, at least in Denmark (Christensen et al., 1995).

Important variables are the time and duration during pregnancy when nutritional deficits were experienced. For example, hyperglycemia *slows* embryo growth early in pregnancy, whereas diabetes of pregnancy at later stages may *accelerate* fetal growth (Barker et al., 1993a). Retarded fetal growth is detected in fetuses of diabetic mothers by 7–14 weeks and reflects the degree of their diabetes (Pedersen and Mølsted-Pedersen, 1981). Moreover, the growth-retarded fetuses had more developmental defects in heart and other organs, which is consistent with increased fluctuating asymmetry in offspring of diabetic monkeys (Kohn and Bennett, 1986).

These influences of low birth weight appear to account for a minor but significant proportion of adult morbidity in most populations. Similarly, a study of fetal malnutrition and adult blood pressure showed extensive overlap among individuals (Langley and Jackson, 1994). Thus, other individual factors may mitigate adversity during development. Provisionally, low birth weight defines a special at-risk subpopulation that may prove to be more sensitive to environmental risk factors for vascular disease such as tobacco, stress, and other lifestyle parameters. Thus, low birth weight alone may not predict adult risk factors for diseases. Many other studies can be added to these examples that indicate the lasting effects on adult pathophysiology that result from the maternal environment (e.g., Desai et al., 1995; Hales et al., 1996; Napoli et al., 1997; Waterland and Garza, 1999).

The extent of developmental programming (Waterland and Garza, 1999), however, is far from clear. For example, children conceived during a terrible famine in rural Finland in 1866–1868 had normal life expectancy (Kannisto et al., 1997). Crops failed for three consecutive years and epidemic diseases increased infant mortality threefold, to 40%. After a good crop in 1868, mortality rates quickly returned to prior values. The detailed records show that the mortality rates and life expectancies of cohorts born just before or after did not differ from the famine years by more than 2%. Data on birth weight, extent of selection in utero, and later causes of death are lacking.

Two other studies of wartime famine will eventually illuminate these important issues. Children born during the blockade of Leningrad (St. Petersburg) in 1941–1944 had mean birth weights about 17% below the normal mean (Stanner et al., 1997). Fifty years later, those exposed to starvation as fetuses or infants did not differ from a nonstarved reference group in Russia in height, weight, or obesity; in glucose tolerance and insulin; or in blood pressure and most clotting factors (but von

Willebrand factor was elevated). However, Dutch children exposed to maternal starvation in mid gestation during the 6 month famine of 1944–1945 are showing impaired glucose tolerance as adults, particularly in an obese subgroup, now middle-aged (Ravelli et al., 1998; Ravelli and van der Meulen, 1993; Kahn et al., 1998). These and other populations with suboptimal nutrition at various times during development are being followed closely for effects on aging and life span, and it is premature to draw conclusions at this time.

4.5. Conclusions

The environment can exert a wide range of different chance effects on the developing individual that influence later outcomes of aging. We showed chance effects from uterine neighbors via sex steroid levels. Variable levels of estrogens from endogenous or exogenous sources strikingly influence the growth of the prostate, and may thereby influence risk of later prostate changes with age. The programming of the HPA axis by neonatal experience in rats raises the possibility that elevated glucocorticoids later in life may have been programmed in part from earlier life. Chance differences in maternal nutrition can also have important consequences, such that diabetic mothers tend to increase the risk of hypertension in their offspring. No information is available on cognitive changes during aging for those who were exposed to toxic levels of ethanol or other stressors during pregnancy. This is a topic of considerable concern, and little is known about the thresholds for inducing developmental abnormalities in the brain and other organs. Thus, for each of us, the expression of our inherited genes is subject to immediate and transgenerational environmental factors that can greatly modify outcomes of aging. Because the types of chance-based phenomena vary so widely in their nature, it is not obvious how to develop a general framework to represent all the kinds of external mechanisms at work upon an individual.

5

Limits of Determinism in Aging

5.1. Recapitulation and synthesis

This inquiry about individual outcomes of aging was motivated by the puzzle posed by the breadth and diversity of the variations in individual life spans and outcomes of aging in inbred laboratory organisms: nematodes, fruitflies, and mice. The puzzle is that these variations are at least as great, relative to the life span of these confined species, as those of human identical twins who experience far more diverse environments during life spans that are 30- to 1000-fold longer. Reproductive aging in mammals gave clues to this puzzle. Individuals differ widely in the numbers of oocytes acquired during development. In turn, individual differences in the pool of irreplaceable oocytes influence the timing of reproductive aging, as shown by partial ovary ablation experiments in mice. Moreover, despite great progress in identifying genetic risk factors for Alzheimer's disease and other conditions of aging, gene carriers vary widely in the time of onset and severity of disease. These diverse phenomena are not readily accounted for in classical quantitative genetics models as direct outcomes of the genotype and its interactions with the *external* environment. We were thus led to examine differences in individuals that arise during development and that might influence outcomes of aging.

We assembled evidence from diverse sources and we posed a new set of questions about the contribution to later aging of chance events during development. There are also broader implications of chance developmental variations for studies on the heritability of human phenotypes other than aging. Although chance variations during development are well known within subfields of developmental biology, there has been little general recognition of their contribution to variations in adult form (cell number and organ structure) and function (organ system physiology, including homeostasis of cell replacement).

We developed the hypothesis that *chance variations in form and function, arising through development, affect individual baseline functions and individual responses to the external environment, and so modify outcomes of aging*. The outcome of an individual's life history rests, metaphorically speaking, on a three-legged stool: variation in each leg (genes, environment, chance) tilts the stool, and each can influence the others.

We recognize a risk that some readers may extrapolate our arguments beyond the context that we have been careful to define. In particular, we see no reason to challenge the great and hard-won truth of modern biology that development is fundamentally controlled by gene action. Multicellular organisms have evolved powerful self-correcting mechanisms, which compensate for intrinsic (local) developmental variations, as well as fluctuations in the external environment. Nonetheless, we must deal with clear evidence that degrees of variation in form and function arise during development and persist in young adults. The extent to which developmental variations, at subcellular and cellular levels, persist into *adult* life has not been systematically characterized, even for *Caenorhabditis*. Much less is known for other laboratory models. Of course, the unexplained variance in studies of human twins or of inbred laboratory animals can have many origins besides intrinsic developmental variations. For instance, it is hard to measure neural functions that are sensitive to the external environment or that fluctuate with the physiological state. We also do not know much about how variations in the external environment interact with ongoing physiological fluctuations in adults.

This closing chapter begins by summarizing main points from the previous chapters. We then consider more generally how development sets the stage for aging and how developmental variations interact with random damage accumulated during aging. We apply these perspectives briefly to current public concerns about human cloning. Last, we outline scientific issues that we anticipate will rise from our platform, including approaches in statistical biology to resolving the influences of adult lifestyle on outcomes of aging. For easier reading, in this chapter we generally do not repeat literature citations from earlier chapters.

Chapter 1 described the wide individual variations in life span in human identical twins and in three laboratory animals: the mouse, the fruitfly, and the nematode (Fig. 1.1). The human twins are strictly isogenic. The self-fertilizing nematode *Caenorhabditis* may be considered isogenic within a tiny margin; the inbred fruitflies and mice, however, may carry a larger margin of undefined genetic variability due to transposable and unstable DNA sequences (Table 1.4, note g). For the nematode, external environmental differences can be ruled out as causes of the twofold differences in life spans, because the worms were grown in liquid culture with constant agitation. Reproductive aging was also shown to vary widely, epitomized by inbred mice, which differ extensively in the time of onset of the lengthening and eventual loss of estrous cycles (Fig. 1.2). These individual variations occur during nearly 50% of the life span. Because reproductive aging in female mammals is largely driven by loss of ovarian oocytes, we developed the hypothesis that individual differences in reproductive aging are direct products of variation in number of oocytes formed during development. We then showed the wide individual variation in oocyte numbers of inbred mice (Fig. 1.3). Variations in oocyte numbers are among many examples of cell number variations in adult organs, which arise by chance during development.

The nervous system also contains variable numbers of postmitotic and irreplaceable neurons (section 1.4.3). However, in inbred rodents, the range of variations in neuron numbers among individuals is smaller than that of oocytes. Genes that determine strain differences in neuron numbers are likely to be identified in the next few years. It will be of interest to learn if these genes also influence mean cell numbers in the ovary and elsewhere and to quantify variance in cell numbers among individuals of the same genotype. A precedent for allele effects on cell number variance is the mouse mutant *tabby*, which increases variability in the number of whiskers (Dun and Fraser, 1959). In grasshoppers, parthenogenic clones differed in the variability of cell body locations in visual system neurons (Goodman, 1978), which implies allelic effects on the precision of cell migration. The genetic influences on the variance of life span observed in inbred mouse strains (de Haan et al., 1998; Table 1.3, footnote c) could be mediated by variations in cell number or cytoarchitectonics which, for some molecular or cellular systems, would, in turn, modify a threshold for pathogenesis (Fig. 1.9, Fig. 2.6).

A different type of variation is the fluctuating asymmetry within individuals (section 1.5.3). Most reports in this fast-growing literature have emphasized the externally measurable features. Beneath the surface is an even greater range of fluctuating asymmetries that arise from the same types of chance variations during development. Brain regions may also differ in relative size between left and right sides, independent of handedness. Clear examples are the hippocampus (Fig. 1.6A) and the planum temporale (Fig. 1.11A; section 1.5.2). Even identical twins have different sizes and shapes of their cortical gyri. Moreover, one pair of identical twins used different sides of their brains for the control of swallowing muscles (Fig. 1.11B). This so far unique example implies that cognitive processes may utilize different neural pathways in identical twins. The lateral line system of larval frogs shows striking fluctuations in neuron number within an individual (Fig.1.10).

A major question is how much the variations in neuron cell number and brain region size determine brain functions. Brain imaging studies of twins may further illuminate this question. The extensive plasticity of neural pathways is already shown in the reorganization of brain fields after loss of appendages. Thus, compensatory processes at more global levels of brain circuit function appear to tolerate local variations in cell number or synapse density. Another indication of this is the remarkable concordance of EEG patterns and amplitude of evoked potentials in identical twins, which implies compensation at a higher level for circuit variations (section 1.5.2). These processes, which integrate the micro- and macroscopic domains of brain function, are as yet poorly understood.

A related question is how variations in cell numbers are corrected or constrained during development. The nematode is an extreme example, with its tightly controlled numbers of cells. However, the positions of cell bodies and the contacts between cells are less precisely determined in certain body regions (section 1.4.3). Further studies are needed to ascertain if these finer variations in body architecture contribute to individual differences in aging. We discussed a model for how individual variations in neuron numbers determine individual thresholds for the extent of loss required before functional deficits are manifested during aging or age-related diseases (Fig. 1.9B).

Cell differentiation is determined by ligand-gated receptor activation, both at the gene level by binding or transcription factors and at the membrane level by signal cascades initiated by binding of external and internal ligands to membrane receptor proteins. In both cases, a large number of molecular collisions by Brownian motion (random walks) are required before a ligand fortuitously collides with and binds to the appropriate receptor site. The stochastic nature of these processes implies a statistical distribution of response times, probably skewed with a long tail. There is evidence that the movement of particles by Brownian motion obeys the laws of mathematical chaos, rather than being strictly random (Durr and Spohn, 1998). However, since chaotic behaviors are characterized by extreme sensitivity to initial conditions (here, the positions and velocities of the particles), and since the initial positions and velocities are themselves largely random, the distinction between true randomness and chaos is immaterial in the present context.

Crucial molecular processes have evolved to cope with chance variations in the molecular diffusion of signals. First, scaffolding proteins organize macromolecular assemblages in subcellular compartments such as the cell nucleus and at cell membranes. The maintenance of a high degree of subcellular organization is required for efficient transfer by diffusion across a distance of signals like cAMP, or for efficient interactions of macromolecules (Ranganathan and Ross, 1997; Nickerson et al., 1995; Mastick et al., 1998). Second, the cell cycle is famous for the carefully regulated checkpoints, such as the G_1/S checkpoint, that impose an orderliness on the cell division process in proliferating tissues, despite the intrinsically variable arrival times of various signals. There are few data yet on the tissue variations of concentrations of signaling molecules (peptides, steroids, etc.) and the size of the various compartments within which they must effectively diffuse. Errors in macromolecular biosynthesis and endogenous oxidative damage to DNA, RNA, and proteins occur all the time, but are, in part, corrected by enzyme and biochemical repair systems. These anti-aging mechanisms evolved particularly in association with the slow maturation and prolonged reproductive schedules that last from years to decades in many vertebrates (Finch, 1990; Kirkwood, 1992).

Chapter 2 considered reproductive aging in mammals. An experimental test of the threshold model (Figs. 1.9, 2.6) showed that the number of oocytes directly determines the age of reproductive senescence (Figs. 2.4 and 2.6). These variations are consistent with the 50% smaller heritability among human twins of the age at menopause than menarche. Oocyte number variations arise early in development, when primordial germ cells proliferate and are eliminated to varying degrees. We do not know if germ cell death during ovarian development involves the same causes as postnatal or postpubertal follicular atresia.

The number of oocytes also sets the risk of several important outcomes of reproductive aging, including Down syndrome and other fetal aneuploidies, and estrogen-sensitive diseases that accelerate after menopause, including heart disease and osteoporosis. Premature increases in fetal aneuploidy occur in mice with partial ovariectomy and in women with Turner syndrome who have congenital ovarian deficits, which indicate that fetal aneuploidy is more an association with the numbers of remaining oocytes than maternal age per se. These findings give a rationale to develop drug interventions to slow the loss of ovarian oocytes and thereby reduce the risk of fetal aneuploidy. However, we do not know if the loss of oocytes itself results from quality assurance mechanisms to eliminate oocytes damaged by intrinsic chance processes.

Because the loss of sex steroids at menopause increases the risk of osteoporotic fractures and heart attacks, chance variations in the oocyte stock will influence the risk schedule for these conditions. The positive effects of estrogen replacement on heart attacks and osteoporosis are becoming better established, and may extend to a lower risk of Alzheimer disease as well. As the use of estrogen replacement continues to increase in some human populations, the negative health impact of an early menopause may be minimized.

Variations during human pregnancy in the fetal blood levels of sex steroids (Fig. 2.12) may also be derived from individual variations in germ cell numbers in male and female fetuses. We hypothesized that differences in steroid levels during development within the physiological range could modify the numbers of steroid-sensitive cells in the reproductive tract and nervous system. Human and rodent brains show considerable overlap in the numbers of neurons in sexually dimorphic regions (Figs. 1.7 and 2.12). Song birds show similar effects, because the amount of testosterone in the egg determines the juvenile social rank (Fig. 2.15).

We also suggested how individual differences in benign prostatic hypertrophy (BPH) during aging in identical twins (Fig. 2.14) could arise from developmental variations in cell numbers. The modest heritability of BPH and prostate size (Section 2.6.2) is paralleled by major individual differences in the size of the prostate region that contributes most to BPH. These differences may arise during development, in possible association with variations in fetal blood estradiol, as suggested by mouse models (Figs. 4.3, 4.5). Prostatic cancers typically arise in a different zone and with no association to BPH, but may be influenced by variable cell–cell interactions. Together, these examples show how several important outcomes of aging in the reproductive system can be traced to chance variations during development.

Chapter 3 examined how variations in the fates of individual cells arise during development and in adult life, and how these variations contribute to the unfolding of individual senescent phenotypes. The variations addressed here are less obvious than variations affecting the numbers of a particular cell type (oocytes, neurons) or the size of particular organs or structures, but they are likely to be pervasive throughout the organism. The basic mechanisms contributing to chance variations in cell fate are asymmetry in cell divisions, randomness in cell differentiation and cell death, and variations in cell migration.

Multicellular organisms comprise complex hierarchies and lineages of cells. They are effectively mosaics of subclones that have differing degrees of relatedness to one another. Thus, a cell may differentiate down alternative pathways according to which pathway is activated first and thus be subject to chance in fate determination. Mammalian tissues can vary in clonal heterogeneity, that is, the incorporation of separate, cross-differentiating cell clones versus a single clone. In the developing cerebral cortex, retroviral marker studies showed that neurons from a single clone can disperse across wide distances and participate in different cytoarchitectonic areas. Other tissues, particularly those with continual renewal from stem cells such as the hematopoietic system and the intestinal epithelium, exhibit a dynamic clonal structure in which the composition of the tissue at any one time results from the expansion of varied, but distinct clonal populations. After clones expand, the cells are eventually discarded or die, so that a continuous process of clonal succession, or attenuation, occurs. Again, more global levels of regulation maintain tissue organization. With advancing age, stem cells may decline in some tissues, resulting in greater chance fluctuations in the clonal composition of the tissues. Skewing of X-chromosome inactivation patterns in blood lymphocytes increases with age in women, suggesting a stochastic clonal loss, possibly with stem cell decline.

Cultured diploid somatic cells show marked variation in clonal division potential (Figs. 3.11, 3.14, 3.15). It is well known that diploid mammalian cell cultures have a finite replicative life span, the "Hayflick limit". But this limit is anything but a fixed constant. Replicate cultures of any diploid cell strain exhibit considerable variation in proliferative potential (section 3.2.5). Moreover, within a mass population individual cells vary extensively in the numbers of cell divisions they can support. The *average* number of divisions of course declines progressively with advancing cell culture age, but even early passage cultures exhibit wide variance in clonal division potential. Furthermore, if a clone is grown from a single cell isolated from the mass population, and separate subclones are then established from this original clone, the resulting subclones soon reveal wide variance in their own proliferative potentials. Likewise, the two daughter cells from a single division vary greatly in their division potential. This example provides incontrovertible evidence that the determination of cell replicative senescence is stochastic, like many other examples of cell fate determination. Immortalized cell lines, such as those derived from malignant tissue, also show variation in clonal growth. Some of the cells, of course, have unlimited division potential, but a sizable fraction (30% in the case of some HeLa sublines) only divide a finite number of times.

A long-standing hypothesis suggests that somatic genome instability—in particular, chance somatic mutation—contributes causally to aging processes. There is now much evidence in mammals for the more-or-less progressive accumulation throughout life of somatic cell DNA mutations, epigenetic defects such as loss of DNA methylation patterns, and abnormalities of chromosomal number and structure. Together, these lesions result in a general drift away from the unique genomic identity of the zygote. The age-related increases of certain cancers, for example, of the breast and prostate, are associated with somatic cell gene mutations. Moroever, there is increasing evidence for dysfunction from DNA instability in some tissues with low cell turnover. Mitochondrial DNA (mtDNA) deletions accumulate strikingly during normal aging in certain brain regions and skeletal muscles, whereas nearby regions may have 1000-fold fewer mtDNA mutations (section 1.7.1). Three observations indicate the importance of both systemic and local cell factors in mtDNA instability: the major tissue differences in mtDNA deletions during normal aging in humans, the observations that accumulation of mtDNA deletions is slowed by food restriction in aging rats, and the evidence that accumulation of mtDNA deletions is greater with hyperglycemia in aging diabetic humans. Despite the now abundant evidence for age-related damage to somatic genes, we are far from a complete inventory of the extent of different types of DNA damage in specific cell types. The evidence suggests a mosaic of aging, with major differences among cell types in the amount and type of damage.

Chapter 4 considers chance features in the external environment of the embryo. In several rodents and in human twins, brain development is influenced by the sex of the neighboring fetus (section 4.1). Adult female rodents show different degrees of masculine behavior, depending on whether the two neighboring fetuses were female (2F) or male (2M). As adults, 2M female mice cease reproducing 2 months early (Fig. 4.4); the cause appears to be hormonal, but the ovarian stock of oocytes has not been characterized. Thus, the random distribution of embryos of different sex has major outcomes to reproductive aging, as well as for risky male-type aggressive behaviors, which would be expected to decrease life expectancy. Moreover, offspring of 2F females had more female offspring themselves.

Humans show modest effects of fetal intrauterine interactions on functions of the inner ear. In about 35% of all adults, the cochlea emits weak tones (otoacoustic emissions, OAE), which are less common in men than in women (Fig. 4.1). Women from opposite-sex twin pairs have a masculinized statistical distribution of OAEs. However, there is no evidence for major effects of fetal interactions with a male co-twin on reproduction.

The external environment as transmitted through the mother can impose chance adversity on the fetus, with impact on adult functions and aging. Maternal grooming in rodents can set the sensitivity of stress responses in neonates, so as to prolong blood glucocorticoid elevations after a stress (section 4.3.2). At later adult ages, hippocampal neuron loss is increased as a consequence of greater glucocorticoid exposure (Fig. 4.9). Developmental stress can also increase fluctuating asymmetry (section 4.3.1). Environmental estrogen studies with mouse models modify prostate development in a direction consistent with increased prostate disease in older men. Data from rodents and humans indicate that, under some circumstances, prenatal nutritional influences may predispose to adult onset heart disease and diabetes (section 4.4). These examples of chance impacts from the external environment on fetal development point to a large realm of developmental influences on outcomes of aging. Moreover, some of these perturbations may impair reproductive fitness of young adults, for example, through the increase of fluctuating asymmetry in external traits that modify mate selection by physical attractiveness.

While we have emphasized throughout the book that there are chance aspects of development, nonetheless we must recognize that many aspects of aging are directly derived from the body plan established during development under genomic authority. Readers may recall the very different canonical patterns of aging in nematodes, fruitflies, and mammals (Table 1.1). Foremost among determinants of aging are the rate and extent of cell and molecule replacement. Adult mammals cannot replace their dwindling stock of ovarian oocytes, whereas new red and white blood cells are constantly formed even at the oldest ages. Adult fruitflies have very few somatic cell types that proliferate, while nematodes have none. In each species, therefore, the capacity for regeneration and repair is set during development as an outcome of differential gene regulation that restricts the inventory of active genes, including those that allow for cell replacement.

At a molecular level, most mRNA and proteins of adult mammals are replaced through ongoing biosynthesis. Intracellular proteins have half-lives ranging from minutes to days. The constant turnover removes damaged molecules. However, DNA in slowly replicating or postmitotic cells is at risk of accumulating various types of chemical damage, which include environmental toxins (Randerath et al., 1993; section 1.7.1). Furthermore, some extracellular proteins are not replaced in adults and therefore can also accumulate damage. Molecules that are particularly long-lived include collagen and elastin of tendon and connective tissue, crystalins of the lens, and dentin in teeth, which may be as old as the individual. We may assume that random damage is constantly occurring to macromolecules, even in embryos. There is no escape from some contact of macromolecules with free radicals, which are an unavoidable by-product of oxidative metabolism. There is also no escape from chemical attack of exposed protein ϵ-amino groups to nonenzymatic glycation from the omnipresent glucose and other sugars. Oxidative damage accrues during aging in every species examined. One cause of the risk of such damage is determined during cell differentiation by setting the molecular turnover rates.

Errors in DNA and RNA synthesis cause mutations and misincorporation, which generally have no immediate consequence to cell functions because of their low frequency. However, if unrepaired, mutations in DNA can persist, which, if combined with further hits acquired "during aging," may allow a clone of cancer cells to emerge. There are also alterations in the pattern of DNA methylation, which influence gene expression, and which may persist in a cell lineage. The loss of methyl groups may lead to inappropriate activation or repression of key genes in differentiated cells, for example, tumor suppressors. The absence of cell replication in most somatic cells of adult fruitflies and its complete absence in nematodes suffice to explain the absence of tumors during aging in these species (Table 1.1).

On the other hand, adult survival is enhanced by the random recombination and mutation of immunoglobulin genes during lymphocyte differentiation, in which a range of base substitutions increases the diversity of the immune response. The selective enhancement of recombination frequency in the variable (V) regions of the immunoglobulin genes serves a specific, adaptive function. However, the recombination and mutation rates are not uniform across the genome—other mutational "hot spots" have been observed. In general, actively transcribed genes are repaired at a higher rate than inactive genes.

The ongoing cell replacement in many tissues of mammals allows for a category of chance outcomes of aging that would not be predicted in nematodes and fruitflies. Given clonal switching during lineage differentiation, individuals will continually evolve further differences in their host defense reactions.

Early vascular lesions are another link between development and aging that extends concepts from earlier chapters. A long-standing hypothesis is that vascular lesions of later life have neonatal precursors: "preatherosclerotic lesions" (Leistikow, 1998). Postmortem studies of human neonates indicate that preatherosclerotic lesions in coronary arteries may be common. The lesions include degrees of stenosis from intimal thickening in association with proliferation of the intimal cell layer and accumulations of foam cells (Hirvonen et al., 1985; Davies, 1990; Stary, 1990, 1994; Leistikow, 1998). Although their incidence in general (nonmorbid) human neonatal populations is unknown, preatherosclerotic lesions with modified smooth muscle cells are frequent in neonatal pigs (Bolande et al., 1995, 1996). Catecholamine-induced acute hypertension, just before or after birth, may intensify preatherosclerotic lesions (Bolande et al., 1995). Acute perinatal hypertension can arise through catecholamine surges, especially during the circulatory transitions at birth, which cause varying degrees of transient anoxia. The acute hypertension and hypoxia in neonates could be idiosyncratic, depending on chance events in the maternal (uteroplacental) or neonatal environment. For example, the second-born twin has lower mean umbilical arterial and venous pO_2 (Chang et al., 1990; Young et al., 1985), which could promote atherosclerotic lesions. It might be fruitful to examine if later clinical heart disease is associated with birth order in co-twin pairs.

Alzheimer disease (AD) may also fit this pattern. In normal humans, scattered neurofibrillary lesions are found by the third decade (Braak and Braak, 1998), but at far lower density than the criterion for AD. The other major postmortem diagnostic criterion of AD is the accumulation of aggregated amyloid β-peptide (Aβ) in senile plaques, which are not found in brains before puberty in any mammal susceptible to these changes (Finch and Sapolsky, 1999). In Down syndrome (chromosome 21 trisomy), most individuals precociously develop AD-type neuropathology in early adult years. However, elevations of sex steroids at puberty are not required. Early deposits of aggregated Aβ are reported in some Down's infants and children (Leverenz and Raskind, 1998; Ter-Minassian et al., 1992), which is very much earlier than the typical 80 years in genetically normal individuals. Another type of amyloid (serum amyloid P, SAP), which is not associated with AD, is common in the skin of children aged 4 years, but not before (Khan and Walker, 1984; Herriot and Walker, 1989). Thus, the aggregation of Aβ and SAP begin relatively early in life (Teller et al., 1996; Pepys et al., 1978).

The early onset of AD-like molecular changes in Down syndrome raises more general questions about gene dosage and the incidence of pathology. The triploid chromosome 21 in Down syndrome carries the gene for APP (amyloid precursor protein) that encodes the APP molecule from which amyloid β peptide (Aβ) is derived. We suggest that spontaneous variations in the expression of APP or other genes on chromsome 21 could likewise yield gene expression levels in range of the Down syndrome gene dose. As further support for this concept, genetically identical nematodes appear to differ in the degree of cellular expression of a stress-inducable gene, *hsp-16* (Tom Johnson, pers. comm.) which increases resistance to environmental stress and toxins. Such individual differences in quantitative gene expression, if stable during the life span, could contribute to individual differences in resistance to disease and mortality risk. The hypothesis of gene dose fluctuations is considered below in relation to the variable age of penetrance of risk and susceptibility factors for AD and other age-related disorders.

In summary, the evidence we have assembled shows the huge potential for chance variations during development to influence outcomes of aging. Chance operates at all levels of biological organization. In ascending order, molecular signals move by random-walk Brownian motion; chance collisions of molecules signal gene activity; chance cell migrations shape the brain; chance variations in cell proliferation influence the numbers of oocytes and neurons; chance at fertilization determines the sex of any neighboring fetus and thereby influences brain development; and, by chance, a developing organism is exposed to adverse conditions in the external environment. The examples selected for discussion anticipate a more comprehensive framework of chance in life history.

5.2. Genetic predictions and their limitations

We now extend our inquiry to consider the predictive value of genetic information, which is also pertinent to ethical concerns. There will soon be extensive data on genetic variations that are risk factors for diseases of aging. We may also anticipate knowledge of genetic variants that are statistically associated with quantitative differences in cognition and behavior, and physiological functions. Our perspectives on the role of chance may temper the concerns that genetic factors strongly determine the outcomes of individual aging.

There is an ongoing cultural transition about genetic information. As humans survive to increasingly advanced ages (Vaupel et al., 1998; Wachter and Finch, 1997; Allard et al., 1998), the role of genetic factors in age-related diseases will gain further attention. Experience with genetic testing for Huntington disease (HD) is a useful precedent for this discussion. HD was one of the first hereditary brain diseases recognized to have a maturity onset, typically between 25 and 50 years (Finch, 1980; Vogel and Motulsky, 1997, p. 132). Carriers develop normally and begin adulthood in full health, typically followed by a devastating loss of motor control and cognitive decline during the next 5–15 years before death.

The HD test protocol itself is relatively simple: DNA from cells in saliva or blood is amplified by polymerase chain reaction (PCR) to estimate the number of tandem trinucleotide repeats $(CAG)_n$ in the *huntingtin* gene (exon 1; Huntington Collaborative Research Group, Macdonald et al., 1993). In numerous family pedigrees, the number of tandem CAG repeats is severalfold larger on average in those who eventually become sick than in unaffected family members (Culjkovic et al., 1997). Lacking prevention or effective treatment, it is not surprising that few family members of HD carriers have chosen to be tested. Genetic indications often bring a tremendous emotional burden (one person committed suicide after learning of the diagnosis by telephone without prior counseling).

However, the genetic test for HD is not always easy to interpret for three reasons. (1) A positive test does not predict the age of onset. (2) There is overlap between normal and HD genotypes, such that some individuals with clinical HD have CAG repeats in the upper normal range; conversely some individuals with repeats in the diagnostic HD range have reached advanced ages in good health (Rubinsztein et al., 1996). Thus, the HD genotype may not show full penetrance, as was widely believed. Difficulties in assigning HD risk are even greater in sporadic cases without family pedigrees to evaluate the pathogenic range of $(CAG)_n$. (3) Flanking DNA sequences to $(CAG)_n$ of the *huntingtin* gene may also contain variant $(CCG)_n$ repeats (Andrew et al., 1994; Vuillaume et al., 1998); without detailed attention that might not be given in mass screening, the test could be misinterpreted. While the burden of self-knowledge of HD and other strongly penetrant genetic defects discourages most from genetic testing, others are choosing to test their offspring early in pregnancy for consideration of elective abortion.

The $\epsilon 4$ allele of apolipoprotein E (*apoE4*, for simplicity) raises further issues. While much less penetrant than HD, *apoE4* is among the most common susceptibility factors for hypercholesterolemia and cardiovascular disease. The allele dose effects may be up to severalfold, but differ importantly by sex, age, and population (Davignon et al., 1988; Weisgraber, 1994; Price et al., 1998; Houlden et al., 1998; Roses, 1997; Kardia et al., 1999). The pathophysiology of the *apoE4* protein is associated with its weaker binding to HDL, with consequent elevations of blood LDL and cholesterol (Weisgraber, 1994). Besides their higher cardiovascular risk, *apoE4* carriers also have severalfold higher risk of Alzheimer disease (AD; Fig. 5.1; Meyer et al., 1998; Roses, 1997). With its statistically strong (if partial) associations for *both* cardiovascular and AD, *apoE4* may be among the most common susceptibility factors for disease-specific mortality after 60 years. In current thinking the *apoE4* allele modulates the time of onset, but not the course of the disease.

Yet, after the age of 85 years, the risk for AD in *apoE4* carriers, as well as in *E2* and *E3*, diminishes sharply (Corder et al., 1996; Meyer et al., 1998; Fig. 5.1). Some centenarian *E4/E4* carriers are mentally normal

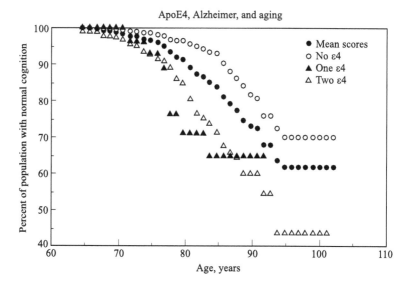

Figure 5.1. The risk for Alzheimer disease in apoE ε4 carriers (*apoE4* in the text) diminishes sharply after 85. Redrawn from Meyer et al. (1998).

(Sobel et al., 1995; Price et al., 1998); however, subtle brain dysfunctions might be shown by sensitive cognitive tests (Morris et al., 1996) or by functional brain imaging (Reiman et al., 1996). Moreover, survival to 100 years means escape from early-onset cardiovascular disease. In any case, the variable penetrance of dementia in *apoE4* carriers reduces its diagnostic value for AD, unless used in combination with clinical criteria (Mayeux et al., 1998; Roses and Saunders, 1997).

The variable penetrance of *apoE4* must be considered with evidence that virtually all humans accumulate some AD-like neuropathology during aging, even if at a modest level. Moreover, AD-like neuropathology appears to be a very general outcome of aging in long-lived mammals. For example, all primates, carnivores, and herbivores so far examined accumulate brain Aβ deposits and neurocytoskeletal abnormalities (Price et al., 1992; Finch and Sapolsky, 1999). Centenarians who appear mentally normal have variable numbers of neurofibrillary tangles (the most recognized neurocytoskeletal abnormality in AD; Mizutani and Shimada, 1992). Similarly, there is a variable occurrence of tangles in young adult brains (Braak and Braak, 1998; section 5.1). The amount of brain Aβ deposits also varies widely and is low in some centenarians (Delaere et al., 1993; Mizutani and Shimada, 1992). These and other data indicate that the brain may function normally, or nearly so, with considerable AD-like molecular changes, even in *apoE4* carriers.

The reduced risk of AD in *apoE4* carriers at advanced ages may indicate interactions with other genes that reduce its effects, as is often the case for susceptibility factors (King et al., 1992, pp. 4–5; Kardia et al., 1999). There are indications of genetic variants that modify neuron numbers during development within a normal range in mice (section 1.4.2). Alternatively, as proposed in chapter 1, chance variations in brain architecture (neuron number and synapse density, cerebral vasculature) could influence thresholds for dysfunction, or the amount of damage that can be sustained above a threshold (Fig. 1.9). Recall from Figure 1.5A the hippocampal size differences among co-twins, including *apoE4* carriers (twins 3, 5, 6). In either case, the presence of the *apoE4* allele may well prove to be subject to other influences that are not currently measured by genetic testing.

At least three genes have rare variants that are associated with familial AD: variants of APP (amyloid precursor protein) and the presenilins, *PS-1* and *PS-2* (100–1000-fold less prevalent than *apoE4* in the general population; Price et al., 1998). These familial AD genes generally act earlier and with higher penetrance than *apoE4* and are recognized as strong risk factors. Nonetheless, the age range of onset can be very wide in presenilin mutations: 34–62 years for *PS-1* (Lopera et al., 1997) and 45–88 years for *PS-2* (Sherrington et al., 1996). AD genes allowing delayed onset of symptoms to ages when there is a high incidence of *sporadic* AD (i.e., AD that arises in the absence of identified familial or genetic risk factors) could be considered as having low penetrance. Death from causes other than AD after 65 years is, of course, common, so some AD gene carriers would in practice be scored as nonpenetrant at death. The issue of "right-censored" data from unrelated mortality is considered in a new multiple-threshold model for AD (Meyer and Breitner, 1998). This situation may well generalize to numerous other genetic risk factors of aging in which the extent of penetrance depends on other genes, on the external environment, or on chance developmental variations that modify thresholds for expression of dysfunction. At this time, most sporadic cases of AD are not associated with any genetic risk factor, which may indicate complex gene–gene or gene–environment interactions (Martin, 1997), as are being resolved for maturity-onset diabetes (see below). Among other candidate genes for AD risk are the variants in the α_1-antichymotrypsin (Wang et al., 1998), VLDL receptor (Okuizumi et al., 1995; Fallin et al., 1997; Chung et al., 1996), α_2-macroglobulin (Liao et al., 1998), and FE65 (Hu et al., 1998). The frontotemporal dementias, which share some pathologic features of AD, have familial forms associated with dominant mutations in the cytoskeletal protein tau (Hong et al., 1998; Goedert et al., 1998), but more common variants have not been associated with effects on brain. The genetics of AD and related disorders are thus extremely complex, with indications of multiple genetic influences of varying intensity and time of impact.

Genetic factors are being sought for many other age-related conditions. A multiplicity of genes that influence the diverse phenotypes of aging was recognized by Martin (1978, 1982), in his prescient analysis of heritable progeroid syndromes, in which certain traits of aging arise precociously in children or in adults. Progeroid genes were estimated to represent about 10% of all the genes associated with human diseases. The exemplar of progeroid syndromes is the Werner syndrome, an adult-onset progeria, caused by inheritance of two copies of a rare (10^{-3}) autosomal recessive mutation (*WRN*), which causes premature graying, intense atherosclerosis, and connective tissue neoplasia, but no AD-like brain changes. Heart attacks commonly cause death before 50 years (Martin, 1997). The *WRN* gene encodes a DNA helicase (Yu et al., 1996; Gray et al., 1997). A polymorphism (DNA change distinct from the WRN mutation) is associated with coronary atherosclerosis (Ye et al., 1997). Other rare progeroid mutations are being examined for variants that modify aging (Martin, 1998).

There are high hopes for identifying many more genetic factors for late-onset diseases through the major efforts to map genes for atherogenesis, cancer, diabetes, and other diseases with midlife or later onset. Current thinking distinguishes genetic *risk* factors with a high degree of penetrance (e.g., Huntington disease or Werner syndrome, the "monogenic" Mendelian diseases associated with mutations in a single gene) from *susceptibility* factors in which effects are conditional, depending on the environment or on other genes (e.g., later-onset Alzheimer disease, maturity-onset diabetes, or vascular disease, each of which is now associated with one or more variants in multiple, distinct genes; King et al., 1992; Vogel and Motulsky, 1997). The variable age of onset in effects of *apoE4* and in the familial *presenilin* and *huntingtin* mutations leads us to anticipate extensive heterogeneity in penetrance of the other genetic factors in later onset diseases, including ages beyond which the risk will diminish. Moreover, for each of the genetic factors with widely variable age of onset, we suggest the potential role of chance developmental variations that modify the threshold for dysfunction (Fig. 1.9).

Another source of heterogeneity of penetrance is *interactions of disease risk alleles* at different loci. For example, in Down syndrome (trisomy 21), *apoE4* on chromosome 19 may accelerate cognitive decline (Del Bo et al., 1997), whereas *apoE2* may be protective (Tyrrell et al., 1998). On the other hand, *apoE4* may protect kidney functions in maturity-onset diabetes (Kimura et al., 1998; Werle et al., 1998). The combinations of several alleles of different genes define "haplotypes," which, as these examples show, are associated with different risk factors for outcomes of aging. We designate alleles that *reduce* the risk of disease and disability during aging as "antigeroid" alleles [after Martin's (1978) concept of "segmental progeroid syndromes"]. In view of allelic interactions among different genes, it may be useful to look for various "antigeroid haplotypes" (combinations of alleles at different gene loci), which will delay and inhibit various age-related disabilities.

In maturity-onset diabetes, the search for genetic risk factors indicates multiple susceptibility genes, with varying impact among populations, and a limited number of dominant familial genes (Morwessel, 1998; Gerich, 1998; Kahn and Rossetti, 1998; Lev-Ran, 1999). As one example of gene–environment interactions, the Pima indians carry an *IFABP* allele associated with the development of obesity with insulin resistance on modern diets (high fat, low complex carbohydrates; Pratley, 1998). Again, a subpopulation of Pimas remain nonobese in the same environment. In this context it is interesting that identical co-twins who are discordant for type 2 diabetes tend to be less obese (Gerich, 1998), which may indicate chance variations during development that influence adipocyte numbers or the neurosecretion of leptin.

Antigeroid haplotypes may also be found in the major histocompatibility complex (referred to as the *MHC* for vertebrates in general and the *HLA* complex in humans). The *MHC* includes about 100 distinct genes and is famous for the unusually large numbers of its allelic variations, which comprise thousands of distinct *MHC* haplotypes with frequencies that differ widely among human populations (Klein, 1986). Besides antigen presentation, *MHC*-encoded proteins also mediate inflammatory responses, the metabolism of steroids and carcinogens, and the action of metabolic and reproductive hormones (Lerner and Finch, 1991; Lerner et al., 1992; Finch and Rose, 1995). In mice, *MHC* haplotypes are associated with differences in female reproductive aging (Lerner et al., 1988). In humans, particular *HLA* haplotypes are associated with rheumatoid arthritis and multiple sclerosis (typical onset, 20–40 years; Klein, 1986), and with type 2 diabetes (typical onset after 40 years; Kasuga et al., 1996; Morwessel, 1998). Other *MHC* haplotypes are associated with longevity in centenarians, with odds ratios in the range of 1.5–2.0. Positive effects of various *HLA* alleles differ between genders and have not been generalized among populations (Ivanova et al., 1998; Akisaka et al., 1997; Rea and Middleton, 1994; other citations in Finch and Tanzi, 1997). Demographic differences in *HLA* haplotype distributions and effects in centenarians resemble those of the far simpler three allele *apoE* ($\epsilon2$, $\epsilon3$, $\epsilon4$) system in heart disease and AD.

Another complexity is that an allele may be protective at one age and hazardous at another, which evolutionary biologists refer to as "antagonistic pleiotropy" (Williams, 1957; Rose, 1991). More generally, there may be wide age group differentials in allele effects on fitness. Candidates for such age group effects might be found in the Pima populations, in which highly efficient storage of reserves would be adaptive in times of food shortages (e.g., for nursing), but might have long-term ill effects manifested as type 2 diabetes, consistent with the proposed existence of such "thrifty" genes (Neel, 1962; Morwessel, 1998). In French centenarians, the D allele of angiotensin-converting enzyme (*ACE*), which is associated with increased risk of cardiovascular disease, was more frequent than the I allele, contrary to expectations (Schächter et al., 1994). One hypothesis is that the D allele has as yet unknown pleiotropic effects and may be protective at older ages. Similarly, genetic predisposition to high blood levels of the low-density lipoprotein (LDL) forms of cholesterol, a common susceptibility factor for cardiovascular disease, might be protective in the 85 + population (Weverling-Rijnsburger et al., 1997).

We briefly recall the evolutionary theory of aging (chapter 1.7.3), which argues against the possibility of just a few pathogenic genes. The force of natural selection diminishes during aging for the simple reason that, in natural populations, most of the reproduction is accomplished by young adults. Even in the absence of aging processes, the natural hazards from accidents, predators, infections, insufficient food, temperature extremes, and so on, will inexorably kill off adults, so only a small fraction could survive to advanced ages. Thus, natural selection must act proportionately more strongly on the younger age groups, which contribute most of the offspring. Consequently, the force of selection is increasingly relaxed at older ages, so the population gene pool will tend to accumulate mutations with delayed adverse effects. This situation predicts a wide diversity of mutations scattered across the genome with delayed adverse effects and also allows for delayed adverse consequences of genes that confer advantage during development or on young adults (antagonistic pleiotropy; see above). Many of the important genes may be those affecting somatic maintenance and repair, as predicted by the disposable soma theory (section 1.7.3). Others may be a heterogeneous assortment of late-acting mutations. As yet, we do not know what those genes may be or the ages of their adverse effects.

Lastly, we return to the instructive discordances of identical twins, which extend to many examples besides life span (Table 1.3), the onset age of AD, and lumbar disk degeneration (Fig. 1.11). Ongoing studies of Swedish twins (Swedish Adoption/Twin Study of Aging) show declining heritability of several physiological traits at later ages; for example, non-shared environmental factors are increasingly important for older twins in regard to systolic blood pressure (Hong et al., 1994), HDL cholesterol and triglycerides (Heller et al., 1994), and vital capacity (lung volume; McClearn et al., 1994). These differences could be entirely due to individual postnatal exposure, or could also interact with initial developmental variations, which are currently modeled as non-shared environment (Eq. 1.1).

In summary, the emerging evidence, as we see it, indicates that numerous later onset diseases are not associated with a single genetic risk or susceptibility factor. Multiple genetic susceptibility or risk factors are being found for diverse late-onset diseases. The extent of their sensitivity to developmental variations and to environmental factors, including lifestyle, while far from clear, is nonetheless consistent with the modest heritability of life spans in identical twins (Table 1.3). The value of genetic testing for various risk factors may be limited without detailed knowledge of environmental interactions. However, the limit on this information may be, as we have argued, that the statistically residual risk factors of aging arise through chance developmental variations that influence the threshold for organ system dysfunction (Fig. 1.9).

Emerging knowledge of genetic influences on age-related diseases could, in an extreme scenario, lead to what might be termed the "eugenics of aging." Such a concept evokes, of course, many ethical questions. Here one may consider not only the likely genetic complexity of aging itself, but also the limits to genetic determinism in aging. Some prospective parents may consider purging their family germ lines of genes like *apoE4* that increase susceptibility to adverse outcomes of aging. While drastic and unethical to some, others may justify such a decision by a current ethic that sanctions the abortion of fetuses with fully penetrant genes like cystic fibrosis, Down syndrome, and Duchenne muscular dystropy. But even without ethical objections, we argue that eugenic aging is not attainable merely by pruning a few genes from the family tree, because it is overwhelmingly clear that no single gene will account for all the phenotypes of human aging. We anticipate that additional progeroid alleles will be found during studies of the numerous minor familial diseases of aging. It may be that most if not all humans carry multiple gene variants, each of which may influence aging to some degree, depending on other genes and on multiple environmental factors. The chance developmental variations in body architecture would still lead to wide individual differences in age of menopause and in health at later ages. The low heritability of life spans (Table 1.3) indicates the limitations of a eugenics approach to aging.

These issues of limited genetic determinism in aging also pertain to well-found public concerns about application of cloning technology to humans. Cloning procedures, in principle, engineer a new form of reproduction, through which the genes of a next generation are derived directly from body cells of an individual. Among the biggest headlines of 1997 was the reported cloning of the sheep Dolly from adult somatic cells (Wilmut et al., 1997). The principle of these techniques was developed decades earlier by experimental embryologists: a nucleus from a diploid cell was transplanted into an enucleated egg (Wolpert et al., 1998). The transplanted egg was transferred to a surrogate mother sheep, where it developed. Eventually, Dolly was proved to be a true clone of her mother's somatic cells by DNA fingerprinting of the highly individual DNA microsatellite patterns (Ashworth et al., 1998). Cloning has since been extended to mice, although the success rate is still much lower with cells derived from adult rather than embryo tissues (Wakayama et al., 1998).

In principle, cloning can be repeated to obtain multigenerational clones, which would be far from the first application of such biotechnology. In agriculture, techniques for cloning were developed thousands of years ago for plants—the glass of Cabernet sauvignon wine that you may enjoy probably came from vines that have been propagated for hundreds of years by asexual cloning (Finch, 1990, p. 229). Of course, a pair of identical twins may be considered as the first two members of a "natural" clone derived from a single set of chromosomes, just as would result from artificial cloning.

But how identical can clones really be? If techiques for artificial cloning were ever applied to humans, the resulting individuals would be genetically the same as natural identical twins. Those who fear the creation of clones of identical humans marching through life, like identical robots, may take heed of the acknowledged individuality among twins (Plomin and Daniels, 1987; McGue and Bouchard, 1998). As described in chapter 1, identical twins who reach middle-age have life spans that differ almost as much as unrelated individuals in the same population (Fig. 1.1D). Their brains are not identical in structural details or size, or in the lateralization of certain functions (Figs. 1.6, 1.7, 1.11). If one twin develops Alzheimer disease, the other lags 4–5 years behind on the average (Fig. 1.12A). Moreover, the random shuffling of immunoglobulin genes during development may also lead to qualitatively different immune responses and disease resistance between co-twins.

But what about cognition? We do not attempt to review the complex and controversial literature of the heritability of behavior, cognition, and measures of intelligence beyond a brief survey. Numerous studies of "general intelligence" (IQ) give estimates of broad-sense heritability in the range of 0.5–0.75 (Bouchard et al., 1990; Devlin et al., 1997; McClearn et al., 1997; Herrnstein and Murray, 1994; McGue and Bouchard, 1998). There are indications that IQ heritability is sensitive to age, although some reports diverge, with increased heritability (McClearn et al., 1997) versus decreased heritability (McGue and Bouchard, 1998). A confound in studies of cognition at middle age or later is the wide range of subtle brain changes that currently would not be identified as pathological, for example, glial activation during middle age (Table 1.2). While there is no doubt of substantial heritability of the IQ measure at all ages, the expression of the genetic component can easily be masked by nonheritable variations, as discussed here.

Other cognitive and behavioral traits show less heritability over a wider range than reported for IQ (McGue and Bouchard, 1998; Alarcon et al., 1998; Pillard and Bailey, 1998). For example, verbal speed, visual memory, and various personality traits show heritabilities in the range of 0.2–0.6. Psychopathology also generally shows less than 50% heritability, for example, twin studies of schizophrenia and of postwar traumatic stress syndrome (McGue and Bouchard, 1998). Future work may clarify the relationship between the heritable factors in cognitive performance and the EEG (section 1.4) and brain system interactions (Tonini and Edelman, 1998). Many chance developmental variations in brain circuit structures and functions (chapters 1, 2, 4) could influence the penetrance of genetic factors. This perspective extends from Plomin and Daniels (1987):

> For personality, psychology, and cognition, behavioral-genetic research converges on the conclusion that most behavioral variability among individuals is environmental in origin . . . . Whatever they may be, these environmental influences make children in the same family as different from one another as are children in different families.

We further argue that the nonshared environment includes chance events in brain development. Thus, despite the greater similarity of identical twins than other siblings or the general population, in our view there is little basis to anticipate that cloning could yield multiple generations of the same individual with identical cognition, feelings, or behaviors.

5.3. Research agenda

5.3.1. Attending to the noise

In collecting material for this book, we have been struck by the extensive but scattered data showing chance variations that arose during development. Developmental biologists are well aware of these variations and recognize them explicitly in their studies of asymmetric cell division and stochastic cell fate determination. Research on cell number regulation may soon show how cell number variations are constrained to differing degrees in nematodes like *Caenorhabditis*, where the variations in somatic cell number are minimal, in contrast with *Pangrellus* and other nematode species in which cell numbers vary more among individuals (section 1.5.3.). The regulation of hematopoietic clonal growth is also a major arena for these studies (section 3.2.4). Researchers of fluctuating asymmetry will soon have a rich mechanistic basis for their discussions, which have hitherto largely attended to external anatomy. However, the biology of somatic *molecular* variations is relatively undocumented, with the exception of the vertebrate immune system.

Brain anatomical differences between identical co-twins (sections 1.4, 1.5) are not much recognized in the literature of fluctuating asymmetry. Much could be learned about individual brain functions from a thorough study of regional correlations in symmetry. Ongoing work on twins with functional imaging during cognitive challenges in real time may show the extent of differences in neural pathways. This possibility is suggested by the different lateralization of swallowing muscle maps in the brains of identical twins (Fig. 1.11).

In addition to research addressing the mechanisms that underlie fluctuating asymmetry and other major chance variations, we suggest the value of examining chance variation at a fine scale in developmental processes. A principle of experimental design is to control variation as much as possible. Nevertheless, and despite stringent environmental control, noise from intrinsic variations is inevitable. All experimenters are concerned from the outset that conclusions may depend upon special, sometimes intangible, circumstances of an experiment and may not be generalizable. We suggest further that the ubiquity of noise merits attention in its own regard as part of basic biological mechanisms in which the responses to intrinsic variations arising through chance may be highly informative. Experiments might be designed very differently if the aim is characterize the *variations* of a biological system, in addition to testing differences in means. At the least, the number of experimental units or repeat observations required to estimate variances with sufficient reliability will generally be larger. In conventional experimental design, the variance or the sampling distribution of the data is most often considered in the power- and sample-size calculations, that is, planning for how many replicates are needed to reduce the standard error of the mean in order to resolve the effects being studied. In this context, the variance is often seen as little more than "noise." But if the aim is to estimate the variance itself, a different approach to experimental design is required. Formulae for the sampling distribution are available (for example, see Armitage and Berry, 1994, pp. 85–88), but are largely unfamiliar outside of the realm of mathematical statistics. Somewhat more familiar are the various statistical tests for equality of variances, which are regularly used to check whether data comply with the uniform-variance requirements of widely used "parametric" statistical tests, such as the Student t-test and ANOVA. These homogeneity of variance tests are often embedded within standard statistical software as a validation device. When experiments are designed to measure and compare sampling distributions themselves, these kinds of variance comparison techniques will play a more central role in the analysis of data from studies designed to explore the range of biological variations.

The disposable soma theory (section 1.7.3) provides a framework to consider how the structure and function of an organism reflect design features for the tolerance of chance variations that arise during development. Natural selection only requires the soma to function during the normative life expectancy in the natural environment. This exposes the limits to genetic specification of events that affect later life. By analogy with engineering, some tolerance is built into the living machinery, which is seen in the chance variations affecting cell numbers and in the random faults affecting molecules and higher order structures. Random faults that are deviations from the genetic program include counting errors in cell numbers, as specified through cell fates; missubstitutions during macromolecular biosynthesis; and accumulated posttranslational modifications through random oxidative damage.

5.3.2. Need for a mathematical framework

In chapter 1, we suggested that the conventional view of individual differences arising from just two sources of variance, the heritable variance V_H and the environmental variance V_E, needs to be extended to include a third term representing intrinsic chance variations, V_C: $V_P = V_H + V_E + V_C$ (eq. 1.2). Although V_E can be partitioned into shared and nonshared environmental components, the latter would not in the usual analysis explicitly represent the intrinsic variations during development for which we introduced the third term V_C.

The experimental challenge in resolving these components is formidable. The experimental designs needed to support such analysis must carefully account for the environmental components such that analysis of variance techniques can estimate the extra variance represented by V_C. This will generally be difficult to do. But we believe that the evidence we have found for a significant role of intrinsic variations in development and aging will, in the appropriate circumstances, justify the effort. The great success of mathematical population genetics has been its ability to develop precise models and predictions based on the framework of classical Mendelian genetics. Mathematical quantitative genetics is a mature science with a formalism and terminology of its own. Nevertheless, many of the problems for which analytic solutions are available are constrained by necessary simplification. Problems involving multiple loci quickly become algebraically intractable, especially for gene interactions. Even in the classical dichotomous gene–environment variance model, the quantitative analysis of environmental variations lags behind. We can expect, therefore, that separation of that component of variance that population geneticists traditionally call "environmental" (i.e., that which does not derive from the genes themselves) into the distinct contributions V_E, from the *real* environment, and V_C will be a major challenge. Thus, there is a need for a new framework in which we anticipate that modeling of cell and molecular processes affecting gene actions will play an increasingly important role.

Other approaches to resolving intrinsic chance variations in epidemiological and experimental studies might change the classical strategies. Outliers should not be automatically discarded, without independent evidence that an observation is at fault. Statistical procedures that are sometimes used to exclude outliers in data sets may be deleting exactly those points that, from the perspective of studying chance variations, are of the greatest interest. All too often, data are reported as describing a statistically significant upward or downward trend with age, when what is often far more striking about the data is the huge amount of scatter! If environmental factors are varied and if the residual variance remains the same, this outcome would strengthen the interpretation that the residual variance represents V_C and not environmental noise. Once the quantitation of the role of intrinsic chance is more explicitly included in the aims of investigations, then new statistical methods should follow.

Already, mathematical modeling and biostatistical analysis are seen as essential tools in understanding biology. The value of mathematical models as predictive and analytic tools for the understanding of complex processes is amply shown in carcinogenesis, chemotaxis, development, immunology, neurophysiology, and wound healing as well as aging. If we are to understand the actions of chance, not only will models play a vital role, but also computer simulations will be necessary to predict the statistical distributions of outcomes. If we have competing hypotheses about the mechanisms generating chance variations, models can generate quantitative predictions about the forms of the statistical distributions. As often mentioned in discussing biological variations, most studies have not directly tested for randomness in the biological variations.

The objectives of a theoretical model should be to sharpen understanding of the biological process or hypothesis, to highlight gaps in current knowledge, and to develop a set of testable predictions. To do this a model must balance realism and simplicity. A fully realistic model that represents all of the molecular details will nearly always be intractable. The most valuable models are often those that are constructed when there are important gaps in our knowledge. They can then be used for "what if" kinds of investigation. Thus, model building can highlight issues that need to be investigated. For example, the McAdams and Arkin (1997) model of stochastic mechanisms in biosynthesis (Figs. 1.16, 1.17) allows for probabilities of interactions between molecular partners in any given unit of time, but does not resolve these probabilities into the underlying random-walk processes that describe the molecular movement within the cell. Other mathematical models that treat the stochastic behaviors and distributions of cell populations are described in chapter 3 (e.g., Kirkwood and Holliday, 1978; Grove et al., 1992; Walsh and Cepko, 1993; Loeffler et al., 1997).

Among the mathematical disciplines allied to the life sciences, there has tended to be a division between the biostatisticians or biometricians (who focus on experimental design and data analysis but less on the modeling of biological processes) versus mathematical/computational biologists or biomathematicians (who focus on modeling, often from a deterministic perspective, but less on statistics). Remarkably, there is even a school of thought among biostatisticians that cautions against becoming overly involved in the biological context of the problem, in order to preserve impartiality in data analysis. We believe that the research agenda required to make real headway in the analysis of chance in development and aging calls for a much more robust union of the complementary methodologies of biostatistics and biomathematics with biology.

To name but a few of many challenging problems: Are there unrecognized and functionally significant patterns in cell number variation? What mathematical models are needed to represent their genesis? How do chance molecular variations affect cell fate? How does the variance in fate of individual cells affect the functional properties of cell populations in vivo? Much work is needed to define the limits of genetic determinism in the range of individual variations at each stage in life history; the relationships of these variations between stages; the extent to which new variations arise at later stages through intrinsic chance events; and the relationship of trait variability to particular selective forces. These fundamental questions should attract increasing numbers of researchers who can build effective interdisciplinary bridges between the relevant branches of biology, mathematics, and statistics.

5.3.3. Applications to medicine

Because exposure to various stressors during development increases fluctuating asymmetry (section 4.3.1), we suggest that there might be interventions to reduce stress effects on developmental anomalies. A precedent for this is the recommendation that pregnant women do not use alcohol, tobacco, or other agents that can have toxic effects on their fetuses. But much more could be considered, once we understand how mild toxic and metabolic challenges modify cell fate statistics. For example, there is greater risk of spina bifida in the offspring of families carrying a common gene variant in methylene tetrahydrofolate reductase, which reduces blood levels of folate. The risk of spina bifida may be reduced by supplements of folate during pregnancy (van der Put et al., 1995; Oakley et al., 1996; Oakley, 1997). Other gene variants are anticipated to emerge from the genome projects that will identify risk factors during development that can be compensated by nutritional and other interventions. Genetic influences on the variance of life spans in inbred mice (de Haan et al., Table 1.3, footnote c) imply the existence of quantitative trait loci (Falconer and Mackay, 1996; Lynch and Walsh, 1998) which modify mortality rate variance through effects on long-term pathogenic processes.

Constitutive epigenetic variations in gene expression may also be important as in the example of *hsp16* in nematodes (section 1.6.2, section 5.1). Other examples may be sought in the stochastics of genomic expression; for example, the transcription start intervals (Fig. 1.17) might be sensitive to mild transients in ATP, which could introduce mild dysynchrony in a complex gene program.

The initial numbers of cells might be evaluated to indicate inidividual risks. For example, there is evidence that fetal aneuploidy is mainly a consequence of a dwindling ovarian oocyte pool (section 2.4). Thus, knowledge of a woman's ovarian oocyte pool might be a basis for devising a personal reproductive plan to reduce the risk of fetal aneuploidy. For example, if a young woman were shown to have a subnormal number of oocytes by some imaging technique, then it might be possible to slow the loss of the remaining oocytes by synthetic peptides that antagonize actions of LH or LHRH. Moreover, the risk of benign prostate hypertrophy (BPH) might be managed in men with relatively larger BPH-prone regions of the prostate, which can be detected by puberty (section 2.6.2). We anticipate many further opportunities for preventive approaches to immune, neural, and vascular system disorders during individual aging which identify chance outcomes of development and differentiation.

Recognition that chance plays an important role in modulating outcomes of aging, in addition to the roles played by genes and external environment, may also help to give insight into the so-called "mortality plateaus" of advanced ages (Curtsinger et al., 1992; Vaupel et al., 1998; Kirkwood, 1999). In humans after about age 90 and in several invertebrates, the Gompertzian exponential increase in age-specific mortality rates eventually slows down and may even decrease at advanced age. Genetic heterogeneity is suspected as a cause of mortality plateaus in humans. Given genetic heterogeneity in a population with respect to the rate at which frailty increase with age, then on average the frailer individuals die first; eventually the population comprises only the most robust subset (Kowald and Kirkwood, 1993a; Vaupel and Carey, 1993; Yashin et al., 1995; Pletcher and Curtsinger, 1998; Vaupel et al., 1998). However, genetic heterogeneity does not explain the mortality plateaus seen in *Drosophila*, because mortality plateaus were similar in genetically homogeneous and heterogeneous stocks (Curtsinger et al., 1992). *Caenorhabditis* also had mortality rate plateaus, although isogenic worms showed less of a plateau than genetically heterogeneous populations (Brooks et al., 1994). A nonexclusive alternative to genetic heterogeneity to explain the mortality rate plateaus is that variance in frailty arises through *intrinsic* chance variations during development and adult life.

The evidence from identical twins underlines the uniqueness of the individual. Despite starting life with identical genomes and growing up usually in the same environment, twins diverge. No doctor would diagnose an illness in one twin by examining the condition of the other! This exaggerated example highlights a difference between clinical and basic science. For the wise clinician, each patient's uniqueness is accounted for in the diagnosis and choice of treatment. The art of medicine lies in synthesis of diverse observations, which often are variable and uncertain. By contrast, the art of science is often seen to lie in choosing problems that allow a high degree of control in the variations. We hope that the present inquiry of the role of chance in the development and life course of the individual provides a platform for joining these perspectives.

5.4. Next steps toward a biology of aging and the individual

As we close this initial inquiry into the role of chance variation in development and aging, we are conscious that we have barely scratched the surface of a huge subject that promises to be immensely intriguing and pertinent to many areas of biology and medicine. We hope that the examples and models we presented will stimulate readers to identify and analyze others from their own field. We are firmly convinced that the pursuit of how chance enters into biological systems at all levels and how genetic systems work to constrain endogenous and exogenous noise will help to avoid overly simplistic concepts of genetic determinism. To prompt further inquiry, we offer the following (far from complete) research agenda.

1. Characterize the individual variations of cell architecture and gene expression throughout life history in isogenic organisms raised under strictly controlled environments. Nematodes and other small invertebrates that can be grown in liquid culture should be highly informative about molecular and cellular changes that arise during development and after hatching, and how they are related to individual differences in aging during their short life spans.

2. Examine the relationships between anatomical and functional variability in neural systems (already a goal of developmental neurobiology): cell numbers, cell–cell contacts, variations in subcellular organelles, and so on, for which the study of synaptic spacing (Hellwig et al., 1994; section 1.4.4) is a good precedent. Many neural systems have sufficient redundancy and global mechanisms to minimize the impact of cell deficits or developmental fluctuations. The huge contrast between the effects of lesions on two catecholamine brain systems (Fig. 1.9) shows the need for further study of thresholds.

3. Characterize extracellular matrix molecular variations and cell–cell signaling variations during development and throughout life.

4. Examine how developmental variations interact with external environmental variations. This underdeveloped subject has the potential for a far-reaching impact.

5. Develop in vivo imaging technologies that identify organ system cell variations at birth, for example, ovarian follicles, prostate zones, which may be a basis for preventive medicine and lifestyle choices.

6. Identify the hormonal and metabolic factors that mediate developmental defects in response to environmental stress and toxins.

7. Identify enzymatic and biochemical factors that modify variability. In view of genes that increase variance of certain traits (section 5.1), transgenic approaches might be used to develop animals with greater variation in cell number in specific organs. The suppression of telomere shortening by increasing telomerase levels in cultured diploid cells (section 4.4) and the influences of diet and diabetes on mitochondrial DNA deletions in vivo (section 1.7.1) are precedents for many other modulations of chance somatic cell damage during aging.

8. Evaluate the functional impact of somatic genome damage during aging. The extent of somatic cell genomic integrity and totipotency has obvious relevance to the engineering of molecular and cellular replacement interventions and other therapies.

References

Abbott DH (1984) Differentiation of sexual behaviour in female marmoset monkeys: effects of neonatal testosterone or a male co-twin. In: *Progress in Brain Research* (ed. De Vries GJ et al.), vol 61, pp 349–358, Elsevier Sciences Publishers, Amsterdam.

Abkowitz JL, Catlin SN, Guttorp P (1996) Evidence that hematopoiesis may be a stochastic process in vivo. *Nature Med* 2: 190–197.

Abkowitz JL, Taboada M, Shelton GH, Catlin SN, Guttorp P, Kiklevich JV (1998) An X chromosome gene regulates hematopoietic stem cell kinetics. *Proc Natl Acad Sci USA* 95: 3862–3866.

Adams MS, Niswander JD (1967) Developmental "noise" and a congenital malformation. *Genet Res Cambridge* 10: 313–317.

Afzelius BA (1996) Inheritance of randomness. *Med Hypoth* 47: 23–26.

Ahmed MM, Lee CT, Oesterling JE (1997) Current trends in the epidemiology of prostate diseases: benign hyperplasia and adenocarcinoma. In: *Prostate: Basic and Clinical Aspects* (ed. Naz RK), pp. 3–28, CRC Press, Boca Raton FL.

Akisaka M, Suzuki M, Inoko H (1997) Molecular genetic studies on DNA polymorphism of the HLA class II genes associated with human longevity. *Tissue Antigens* 50: 489–493.

Alarcon M, Plomin R, Fulker DW, Corley R, DeFries JC (1998) Multivariate path analysis of specific cognitive abilities data at 12 years of age in the Colorado Adoption Project. *Behav Genet* 28: 255–264.

Albertson DG, Thomson JN (1976) The pharynx of *Caenorhabditis elegans*. *Philos Trans R Soc Lond [Biol]* 275: 299–325.

Allard M, Lèbre V, Robine J-M (1998) *Jeanne Calment. From Van Gogh's Time to Ours. 122 Extraordinary Years* (trans. Beth Coupland), WH Freeman, New York.

Altman J, Bayer SA (1990a) Mosaic organization of the hippocampal neuro-epithelium and the multiple germinal sources of dentate granule cells. *J Comp Neurol* 301: 325–342.

Altman J, Bayer SA (1990b) Prolonged sojourn of developing pyramidal cells in the intermediate zone of the hippocampus and their settling in the stratum pyramidale. *J Comp Neurol* 301: 343–364.

Ames BN, Saul RL, Schwiers E, Adelman R, Cathcart R (1985) Oxidative damage as related to aging and cancer: the assay of thymine glycol, thymidine glycol, and hydroxymethyl uracil in human and rat urine. In: *Molecular Biology of Aging* (ed., Sohal RS, Birnbaum LS, Cutler RG), pp 137–144, Raven Press, New York.

Anderson AC (ed.) (1970) *The Beagle as an Experimental Dog*. Ames: Iowa State Univ Press.

Anderson RP, Menninger JR (1986) The accuracy of RNA synthesis. In: *Accuracy in Molecular Processes* (ed. Kirkwood TBL, Rosenberger RF, Galas DJ), pp 159–189, Chapman and Hall, London.

Andrew SE, Goldberg YP, Theilmann J, Zeisler J, Hayden MR (1994) A CCG repeat polymorphism adjacent to the CAG repeat in the Huntington disease gene: implications for diagnostic accuracy and predictive testing. *Hum Mol Genet* 3: 65–67.

Anisman H, Zaharia MD, Meaney MJ, Merali Z (1998) Do early-life events permanently alter behavioral and hormonal responses to stressors? *Int J Dev Neurosci* 16: 149–164.

Anokhin A, Steinlein O, Fischer C, Mao Y, Vogt P, Schalt E, Vogel F (1992) A genetic study of the human low-voltage electroencephalogram. *Hum Genet* 90: 99–112.

Apfeld J, Kenyon C (1998) Cell nonautonomy of *C. elegans* daf-2 function in the regulation of diapause and life span. *Cell* 95: 199–210.

Arden NK, Baker J, Hogg C, Baan K, Spector TD (1996) The heritability of bone mineral density, ultrasound of the calcaneus and hip axis length: a study of postmenopausal twins. *J Bone Miner Res* 11: 530–534.

Arkin A, Ross J, McAdams HH (1998) Stochastic kinetic analysis of developmental pathway bifurcation in phage lambda-infected *Escherichia coli* cells. *Genetics* 149: 633–648.

Armitage P, Berry G (1994) *Statistical Methods in Medical Research*, 3rd edn., Blackwell Science, Oxford.

Arnold AP (1990) The passerine bird song system as a model in neuroendocrine research. *J Exp Zool [Suppl]* 4: 22–30.

Arnone MI, Davidson EH (1997) The hardwiring of development: organization and function of genomic regulatory systems. *Development* 124: 1851–1864.

Ashworth D, Bishop M, Campbell K, Colman A, Kind A, Schnieke A, Blott S, Griffin H, Haley C, McWhir J, Wilmut I (1998) DNA microsatellite analysis of Dolly. *Nature* 394: 329.

Aspnes LE, Lee CM, Weindruch R, Chung SS, Roecker EB, Aiken JM (1997) Caloric restriction reduces fiber loss and mitochondrial abnormalities in aged rat muscle. *FASEB J* 11: 573–581.

Austad SN (1993) Retarded senescence in an insular population of Virginia opossums (*Didelphis virginiana*). *J Zool* 229: 695–708.

Austad, SN (1997) Comparative aging and life histories in mammals. *Exp Gerontol* 32: 23–38.

Avdi M, Driancourt MA (1997) Influence of sex ratio during multiple pregnancies on productive and reproductive parameters of lambs and ewes. *Reprod Nutr Dev* 37: 21–27.

Bae E, Cook KR, Geyer PK, Nagoshi RN (1994) Molecular characterization of ovarian tumors in *Drosophila*. *Mech Dev* 47: 151–164.

Baish JW, Jain RK (1998) Cancer, angiogenesis, and fractals. *Nature Med* 4: 984.

Baker CVH, Sharpe CR, Torpey NP, Heasman J, Cylie CC (1995) A *Xenopus* c-kit-related receptor tyrosine kinase expressed in migrating stem cells of the lateral line system. *Mech Dev* 50: 217–228.

Baker CVH, Bronner-Fraser M, Le Douarin NM, Teillet M-A (1997) Early- and late-migrating cranial neural crest cell populations have equivalent developmental potential in vivo. *Development* 124: 3077–3087.

Baker TG (1963) A quantitative and cytological study of germ cells in human ovaries. *Proc R Soc London [Biol]* 158: 417–433.

Bang AG, Bailey AM, Posakony JW (1995) *Hairless* promotes stable commitment to the sensory organ precursor cell fate by negatively regulating the activity of notch signaling pathway. *Devel Biol* 172: 479–494.

Banerjee PP, Banerjee S, Lai JM, Strandberg JD, Zirkin BR, Brown TR (1998) Age-dependent and lobe-specific spontaneous hyperplasia in the brown Norway rat prostate. *Biol Reprod* 59: 1163–1170.

Barker DJP (1994) *Mothers, Babies and Disease in Later Life*. London: BMJ Publishing Group.

Barker DJP (1995) Fetal origins of coronary heart disease and stroke: evolutionary implications. In: *Evolution in Health and Disease* (ed. Stearns SC), pp 246–250, Oxford University Press, New York.

Barker DJP (1999) Early growth and cardiovascular disease. *Arch Dis Child* 80: 305–307.

Barker DJP, Gluckman PD, Godfrey KM, Harding JE, Owens JA, Robinson JS (1993) Fetal nutrition and cardiovascular disease in adult life. *Lancet* 341: 938–941.

Barrett-Connor E, Bush TL (1991) Estrogen and coronary heart disease in women. *JAMA* 265: 1861–1867.

Bartley AL, Jones DW, Weinberger DR (1997) Genetic variability of human brain size and cortical gyral patterns. *Brain* 120: 257–269.

Baskerville A, Cook RW, Dennis MJ, Cranage MP, Greenaway PJ (1992) Pathological changes in the reproductive tract of male rhesus monkeys associated with age and simian AIDS. *J Comp Pathol* 107: 49–57.

Battié MC, Videman T, Gibbons LE, Fisher LD, Manninen H, Gill K (1995) 1995 Volvo Award in Clinical Sciences: determinants of lumbar disc degeneration. A study relating lifetime exposures and magnetic resonance imaging findings in identical twins. *Spine* 20: 2601–2612.

Baunack E, Gärtner K, Schneider B (1986) Is the "environmental" component of the phenotypic variability in inbred mice influenced by the cytoplasm of the egg? *J Vet Med Assoc* 33: 641–646.

Baunack E, Gärtner K, Werner I (1988) Ovary transplantation—a means of estimating prenatal maternal effects in mice. *Physiol Pathol der Fortpflanzung* 4: 235–239.

Bayer SA (1980) Development of the hippocampal region in the rat. 1. Neurogenesis examined with ^{3}H-thymidine autoradiography. *J Comp Neurol* 190: 87–114.

Beale CM, Collins P (1996) The menopause and the cardiovascular system. *Bailliere's Clin Obst Gynaecol* 10: 483–513.

Beckman KB, Ames BN (1997) Oxidative decay of DNA. *J Biol Chem* 272: 19633–19636.

Beckman KB, Ames BN (1998) The free radical theory of aging matures. *Physiol Rev* 78: 547–581.

Behringer RR (1995) The müllerian inhibitor and mammalian sexual development. *Phil Trans R Soc London [Biol]* 350: 285–289.

Berenbaum SA (1999) Effects of early androgens on sex-typed activities and interests in adolescents with congenital adrenal hyperplasia. *Horm Behav* 35: 102–110.

Berenbaum SA, Denburg SD (1995) Evaluating the empirical support for the role of testosterone in the Geschwind-Behan-Galaburda model of cerebral lateralization: commentary on Bryden, McManus, and Bulman-Fleming. *Brain Cognit* 27: 79–83.

Berenbaum SA, Korman K, Leveroni C (1995) Early hormones and sex differences in cognitive abilities. *Learn Indiv Diff* 7: 303–321.

Bergeron L, Perez GI, Macdonald G, Shi L, Sun Y, Jurisicova A, Varmuza S, Latham KE, Flaws JA, Salter JC, Hara H, Moskowitz MA, Li E, Greenberg A, Tilly JL, Yuan J (1998) Defects in regulation of apoptosis in caspase-2–deficient mice. *Genes Dev* 12: 1304–1314.

Bernheimer H, Birkmayer W, Hornykiewicz O, Jellinger K, Seitelberger F (1973) Brain dopamine and the syndromes of Parkinson and Huntington: Clinical, morphological, and neurochemical correlations. *J Neurol Sci* 20: 415–455.

Berry LW, Westlund B, Schedl T (1997) Germ-line tumor formation caused by activation of glp-1, a *Caenorhabditis elegans* member of the Notch family of receptors. *Development* 124: 925–926.

Bibbo M, Gill WB, Azizi F, Blough R, Fang VS, Rosenfield RL, Schumacher GF, Sleeper K, Sonek MG, Wied GL (1977) Follow-up study of male and female offspring of DES-exposed mothers. *Obstet Gynecol* 49: 1–8.

Bigler ED, Blatter DD, Anderson CV, Johnson SC, Gale SD, Hopkins RO, Burnett B (1997) Hippocampal volume in normal aging and traumatic brain injury. *Am J Neuroradiol* 18: 11–23.

Bines J, Oleske DM, Cobleigh MA (1996) Ovarian function in premenopausal women treated with adjuvant chemotherapy for breast cancer. *J Clin Oncol* 14: 1718–1729.

Biondi A, Nogueira H, Dormont D, Duyme M, Hasboun D, Zouaoui A, Chantome M, Marsault C (1998) Are the brains of monozygotic twins similar? A three-dimensional MR study. *Am J Neuroradiol* 19: 1361–1367.

Bix M, Locksley RM (1998) Independent and epigenetic regulation of the interleukin-4 alleles in CD4(+) T cells. *Science* 281: 1352–1354.

Block E (1952) Quantitative morphological investigations of the follicular system in women. *Acta Anat* 14: 108–123.

Block E (1953) Quantitative morphological investigation of the follicular system in newborn female infants. *Acta Anat* 17: 201–206.

Blomquist E, Westermark B, Ponten J (1980) Ageing of human glial cells in culture: increase in the fraction of non-dividers as demonstrated by a mini-cloning technique. *Mech Ageing Dev* 12: 173–182.

Bodnar AG, Ouellette M, Frolkis M, Holt SE, Chiu CP, Morin GB, Harley CB, Shay JW, Lichtsteiner S, Wright WE (1998) Extension of life-span by introduction of telomerase into normal human cells. *Science* 279: 349–352.

Boerrigter METI, Dolle MET, Martus HJ, Gossen JA, Vijg J (1995) Plasmid-based transgenic mouse model for studying in vivo mutations. *Nature* 377: 657–659.

Bohr VA, Smith CA, Okumoto DS, Hanawalt PC (1985) DNA repair in an active gene: removal of pyrimidine dimers from the DHFR gene of CHO cells is much more efficient than in the genome overall. *Cell* 40: 359–369.

Bolande RP, Leistikow EA, Wartmann FS III, Louis TM (1995) The effects of acute norepinephrine-induced hypertension on the coronary arteries of newborn piglets. *Exp Mol Pathol* 63: 87–100.

Bolande RP, Leistikow EA, Wartman FS III, Louis TM (1996) The development of preatherosclerotic coronary artery lesions in perinatal piglets. *Biol Neonate* 69: 109–118.

Bolanowski MA, Russell RL, Jacobson LA (1981) Quantitative measures of aging in the nematode *Caenorhabditis elegans*. 1. Population and longitudinal studies of two behavioral parameters. *Mech Ageing Dev* 15: 279–295.

Boomsma D, Anokhih A, de Geus E (1997) Genetics of electrophysiology: linking genes, brain and behavior. *Cur Direct Psychol Sci* 6: 106–110.

Bouchard TJ, Lykken DT, McGui M, Segal NL, Tellegen A (1990) Sources of human psychological differences: the Minnesota study of twins reared apart. *Science* 250: 223–228.

Braak H, Braak E (1998) Alzheimer's disease starts in early adulthood. *Proc 6th Int Conf Alzheimer's Dis Rel Disord*, 18–23 July 1998, Amsterdam, The Netherlands. [Abstract]

Bridges CB (1927) The relation of the age of the female to crossing over in the third chromosome of *Drosophila melanogaster*. *J Gen Physiol* 81: 689–701.

Bronson FH, Desjardins C (1986) Physical and metabolic correlates of sexual activity in aged male mice. *Am J Physiol* 250: R665–R675.

Brook JD, Gosden RG, Chandley AC (1984) Maternal aging and aneuploid embryos: evidence from the mouse that biological and not chronological age is the important influence. *Hum Genet* 66: 41–45.

Brooks A, Johnson TE (1991) Genetic specification of life span and self-fertility in recombinant-inbred strains of *Caenorhabditis elegans*. *Heredity* 67: 19–26.

Brooks A, Lithgow GJ, Johnson TE (1994) Mortality rates in a genetically heterogeneous population of *Caenorhabditis elegans*. *Science* 263: 668–671.

Bryden MP, McManus IC, Bulman-Fleming MB (1995) GBG, hormones, genes, and anomalous dominance: a reply to commentaries. *Brain Cognit* 27: 94–97.

Bucheton A (1978) Non-Mendelian female sterility in *Drosophila melanogaster*: influence of ageing and thermic treatments. I. Evidence for a partly inheritable effect of these two factors. *Heredity* 41: 357–369.

Buescher M, Yeo SL, Udolph G, Zavortink M, Yang X, Tear G, Chia W (1998) Binary sibling neuronal cell fate decisions in the *Drosophila* embryonic central nervous system are nonstochastic and require inscuteable-mediated asymmetry of ganglion mother cells. *Genes Dev* 12: 1858–1870.

Buehr M (1997) The primordial germ cells of mammals: some current perspectives. *Exp Cell Res* 232: 194–207.

Bulmer MG (1970) *The Biology of Twinning in Man*, Clarendon Press, Oxford.

Burne BD (1958) *The Mammalian Cerebral Cortex* (ed. Arnold E), *Monogr Physiol Soc* 5, London.

Burnet FM (1974) *Intrinsic Mutagenesis: A Genetic Approach to Aging*, Wiley, New York.

Byne W (1998) The medial preoptic and anterior hypothalamic regions of the rhesus monkey: cytoarchitectonic comparison with the human and evidence for sexual dimorphism. *Brain Res* 793: 346–350.

Campisi J (1997) The biology of replicative senescence. *Eur J Cancer* 33: 703–709.

Carey JR, Liedo P, Muller HG, Wang JL, Vaupel JW (1998) A simple graphical technique for displaying individual fertility data and cohort survival: case study of 1000 Mediterranean fruit fly females. *Functional Ecol* 12: 359–363.

Casaccia-Bonnefil P, Carter BD, Dobrowsky RT, Chao MV (1996) Death of oligodendrocytes mediated by the interaction of nerve growth factor with its receptor p75. *Nature* 383: 716–719.

Cauley JA, Seeley DG, Browner WS, Ensrud K, Kuller LH, Lipschutz RC, Hulley SB (1997) Estrogen replacement therapy and mortality among older women. The study of osteoporotic fractures. *Arch Intern Med* 157: 2181–2187.

Caviness VS, Meyer J, Makris N, Kennedy DN (1996) MRI-based topographic parcellation of human neocortex: an anatomically specified method with an estimate of reliability. *J Cogn Neurosci* 8: 566–587.

Cernuda-Cernuda R, García-Fernández JM (1996) Structural diversity of the ordinary and specialized lateral line organs. *Microscopy Res Techn* 34: 302–312.

Chang TH, Jeng CJ, Lan CC (1990) The effect of birth order in twins on fetal umbilical blood gas and apgar score. *Chung Hua I Hsueh Tsa Chih (Taipei)* 46: 156–160.

Charlesworth B (1994) *Evolution in Age-Structured Populations*, 2nd ed., Cambridge University Press, Cambridge.

Chen CS, Mrksich M, Huang S, Whitesides GM, Ingber D (1997) Geometric control of cell life and death. *Science* 276: 1425–1428.

Chen J, Astle CM, Harrison DE (1999) Development and aging of primitive hematopoietic stem cells in BALB/cBy mice. *Exp Hematol* 1–8.

Chenn A, McConnell SK (1995) Cleavage orientation and the asymmetric inheritance of *Notch*—immunoreactivity in mammalian neurogenesis. *Cell* 82: 631–641.

Chippendale AK, Palmer AR (1993) Persistence of subtle departures from symmetry over multiple molts in individual brachyuran crabs: relevance to developmental stability. *Genetica* 89: 185–199.

Christensen K, Vaupel JW, Holm NV, Yashin AI (1995) Mortality among twins after age 6: fetal origins hypothesis versus twin method. *Brit Med J* 310: 432–436.

Chung H, Roberts CT, Greenberg S, Rebeck GW, Christie R, Wallace R, Jacob HJ, Hyman BT (1996) Lack of association of trinucleotide repeat polymorphisms in very-low-density lipoprotein receptor gene with Alzheimer's disease. *Ann Neurol* 39: 800–803.

Claas B, Münz H (1996) Analysis of surface wave direction by the lateral line system of *Xenopus*: source localization before and after inactivation of different parts of the lateral line. *J Comp Physiol A* 178: 253–268.

Clark AS, Goldman-Rakic PS (1989) Gonadal hormones influence the emergence of cortical function in nonhuman primates. *Behav Neurosci* 103: 1287–1295.

Clark MM, Galef BG Jr (1995) A gerbil dam's fetal intrauterine position affects the sex ratios of litters she gestates. *Physiol Behav* 57: 297–299.

Clark MM, Crews D, Galef BG (1991) Concentrations of sex steroid hormones in pregnant and fetal Mongolian gerbils. *Physiol Behav* 49: 239–243.

Clark MM, Robertson RK, Galef BG Jr (1996) Effects of perinatal testosterone on handedness of gerbils: support for part of the Geschwind-Galaburda hypothesis. *Behav Neurosci* 110: 413–417.

Clarke PGH (1990) Developmental cell death: morphological diversity and multiple mechanisms. *Anat Embryol* 181: 195–213.

Clarke PGH, Cowan WM (1975) Ectopic neurons and aberrant connections during neural development. *Proc Nat Acad Sci USA* 72: 4455–4458.

Clarke PGH, Cowan WM (1976) The development of the isthmo-optic tract in the chick, with special reference to the occurrence and correction of developmental errors in the location and connections of isthmo-optic neurons. *J Comp Neurol* 167: 143–164.

Clarke PGH, Rogers LA, Cowan WM (1976) The time of origin and the pattern of survival of neurons in the isthmo-optic nucleus of the chick. *J Comp Neurol* 167: 125–142.

Cohen I, Speroff L (1991) Premature ovarian failure: update. *Obstet Gynecol Surv* 46: 156–162.

Collignon J, Varlet I, Robertson EJ (1996) Relationship between asymmetric nodal expression and the direction of embryo turning. *Nature* 381: 158–161.

Cooke J (1975) Control of somite number during morphogenesis of a vertebrate *Xenopus laevis*. *Nature* 254: 196–199.

Cooke J (1998) A gene that resuscitates a theory—somitogenesis and a molecular oscillator. *Trends Genet* 14: 85–88.

Cooper C (1997) The crippling consequences of fractures and their impact on quality of life. *Am J Med* 103: 12S–17S.

Corder EH, Basun H, Lannfelt L, Viitanen M, Winblad B (1996) Attenuation of apolipoprotein E epsilon4 allele gene dose in late age. *Lancet* 347: 542.

Corfas G, Dudai Y (1991) Morphology of a sensory neuron in *Drosophila* is abnormal in memory mutants and changes during aging. *Proc Natl Acad Sci USA* 88: 7252–7256.

Coucouvanis EC, Sherwood SW, Carswell-Crumpton C, Spack EG, Jones PP (1993) Evidence that the mechanism of prenatal germ cell death in the mouse is apoptosis. *Exp Cell Res* 209: 238–247.

Coucouvanis EC, Martin GR, Nadeau JH (1995) Genetic approaches for studying programmed cell death during development of the laboratory mouse. *Methods Cell Biol* 46: 387–440.

Counter CM, Hahn WC, Wei W, Caddle SD, Beijersbergen RL, Lansdorp PM, Sedivy JM, Weinberg RA (1998) Dissociation among in vitro telomerase activity, telomere maintenance, and cellular immortalization. *Proc Natl Acad Sci USA* 95: 14723–14728.

Cowell LG, Crowder LB, Kepler TB (1998) Density-dependent prenatal androgen exposure as an endogenous mechanism for the generation of cycles in small mammal populations. *J Theor Biol* 190: 93–106.

Crandall JE, Herrup K (1990) Patterns of cell lineage in the cerebral cortex reveal evidence for developmental boundaries. *Exp Neurol* 109: 131–139.

Cristofalo VJ, Pignolo RJ (1996) Molecular markers of senescence in fibroblast-like cultures. *Exp Gerontol* 31: 111–123.

Cristofalo VJ, Allen RG, Pignolo RJ, Martin BG, Beck JC (1998) Relationship between donor age and the replicative lifespan of human cells in culture: a reevaluation. *Proc Natl Acad Sci USA* 95: 10614–10619.

Culjkovic B, Ruzdijic S, Rakic L, Romac S (1997) Improved polymerase chain reaction conditions for quick diagnostics of Huntington disease. *Brain Res Brain Res Protoc* 2: 44–46.

Curtsinger JW, Fukui HH, Townsend DR, Vaupel JW (1992) Demography of genotypes: failure of the limited lifespan paradigm in *Drosophila melanogaster*. *Science* 258: 461–463.

Cutting AE, Höog C, Calzone FJ, Britten RJ, Davidson EH (1990) Rare maternal mRNAs code for regulatory proteins that control lineage-specific gene expression in the sea urchin embryo. *Proc Natl Acad Sci USA* 87: 7953–7957.

Dahri S, Snoeck A, Reusens-Billen B, Remacle C, Hoet JJ (1991) Islet function in offspring of mothers on low-protein diet during gestation. *Diabetes* 40 (Suppl 2): 115–120.

Danley KL, Richardson JL, Bernstein L, Langholz B, Ross RK (1995) Prostate cancer: trends in mortality and stage-specific incidence rates by racial/ethnic group in Los Angeles County, California (United States). *Cancer Causes Control* 6: 492–498.

Darwin C (1868) *The Variation of Animals and Plants under Domestication* (vol II). John Murray: London, first edition; reprinted by Impression Anastalique, Brusells, Belgium (1969).

Davidson EH (1986) *Gene Activity in Early Development*, 3rd ed., Academic Press, New York. p. 67.

Davidson EH (1990) How embryos work: a comparative view of diverse modes of cell fate specification. *Development* 108: 365–389.

Davies H (1990) Atherogenesis and the coronary arteries of childhood. *Int J Cardiol* 28: 283–292.

Davignon J, Gregg RE, Sing CF (1988) Apolipoprotein E polymorphism and atherosclerosis. *Arteriosclerosis* 8: 1–21.

Davis KD, Kiss ZH, Luo L, Tasker RR, Lozano AM, Dostrovsky JO (1998) Phantom sensations generated by thalamic microstimulation. *Nature* 391: 385–387.

Dawson DR, Killackey HP (1987) The organization and mutability of the forepaw and hindpaw representations in the somatosensory cortex of the neonatal rat. *J Comp Neurol* 256: 246–256.

Day JC, Koehl M, Deroche V, Le Moal M, Maccari S (1998) Prenatal stress enhances stress- and corticotropin-releasing factor–induced stimulation of hippocampal acetylcholine release in adult rats. *J Neurosci* 18: 1886–1892.

De Boer RJ, Noest AJ (1998) T Cell renewal rates, telomerase, and telomere length shortening. *J Immunol* 160: 5832–5837.

de Celis JF, Bray S (1997) Feed-back mechanisms affecting Notch activation at the dorsoventral boundary in the *Drosophila* wing. *Development* 124: 3241–3251.

DeCherney AH, Cholst I, Naftolin F (1981) Structure and function of the fallopian tubes following exposure to diethylstilbestrol (DES) during gestation. *Fertil Steril* 36: 741–715.

de Haan G, Gelman R, Watson A, Yunis E, Van Zant G (1998) A putative gene causes variability in lifespan among genotypically identical mice. *Nat Genet* 19: 114–116.

Delaere P, He Y, Fayet G, Duyckaerts C, Hauw JJ (1993) Beta A4 deposits are constant in the brain of the oldest old: an immunocytochemical study of 20 French centenarians. *Neurobiol Aging* 14: 191–194.

Del Bo R, Comi GP, Bresolin N, Castelli E, Conti E, Degiuli A, Ausenda CD, Scarlato G (1997) The apolipoprotein E epsilon 4 allele causes a faster decline of cognitive performances in Down's syndrome subjects. *J Neurol Sci* 145: 87–91.

Delrue C, Deleplanque B, Rouge-Pont F, Vitiello S, Neveu PJ (1994) Brain monoaminergic, neuroendocrine, and immune responses to an immune challenge in relation to brain and behavioral lateralization. *Brain Behav Immun* 8: 137–152.

Desai M, Crowther MJ, Ozanne SE, Lucas A, Hales CN (1995) Adult glucose and lipid metabolism may be programmed during fetal life. *Biochem Soc Trans* 23: 331–335.

Deshner EE, Godbold J, Lynch HT (1988) Rectal epithelial cell proliferation in a group of young adults. Influence of age and genetic risk for colon cancer. *Cancer* 61: 2286–2290.

Devlin B, Daniels M, Roeder K (1997) The heritability of IQ. *Nature* 388: 468–471.

Diamond JM (1982) Big-bang reproduction and aging in male marsupial mice. *Nature* 321: 565–567.

Diamond JM (1986) Why do disused proteins become genetically lost or repressed? *Nature* 321: 565–567.

Diamond TH, Thornley SW, Sekel R, Smerdeley P (1997) Hip fracture in elderly men: prognostic factors and outcomes. *Med J Austral* 167: 412–415.

Dimri GP, Lee XH, Basile G, Acosta M, Scott C, Roskelley C, Medrano EE, Linskens M, Rubelj I, Pereira-Smith O, Peacocke M, Campisi J (1995) A biomarker that identifies senescent human cells in culture and in aging skin in vivo. *Proc Natl Acad Sci USA* 92: 9363–9367.

Do KA, Treloar SA, Pandeya N, Purdie D, Green AC, Heath AC, Martin NG (1998) Predictive factors of age at menopause in a large Australian twin study. *Hum Biol* 70: 1073–1091.

Do KA, Broom BM, Kuhnert P, Todorov AA, Duffy D, Treloar SA, Martin NG. Genetic analysis of the age at menopause by using estimating equations and Bayesian random effects models. *Stat Med*, in press.

Dodson RE, Gorski RA (1993) Testosterone propionate administration prevents the loss of neurons within the central part of the medial preoptic nucleus. *J Neurobiol* 24: 80–88.

Döhler KD, Coquelin A, Davis F, Hines M, Shryne JE, Gorski RA (1984) Pre- and postnatal influence of testosterone propionate and diethylstilbestrol on differentiation of the sexually dimorphic nucleus of the preoptic area in male and female rats. *Brain Res* 302: 291–295.

Dollé MET, Martus HJ, Gossen JA, Boerrigter METI, Vijg J (1996) Evaluation of a plasmid-based transgenic mouse model for detecting in vivo mutations. *Mutagenesis* 11: 111–118.

Dollé MET, Giese H, Hopkins GL, Martus H-J, Jeffrey MH, Vijg J (1997) Rapid accumulation of genome rearrangements in liver but not in brain of old mice. *Nature Genet* 17: 431–434.

DonCarlos LL (1996) Developmental profile and regulation of estrogen receptor (ER) mRNA expression in the preoptic area of prenatal rats. *Brain Res Dev Brain Res* 94: 224–233.

Drickamer LC, Arthur RD, Rosenthal TL (1997) Conception failure in swine: importance of the sex ratio of a female's birth litter and tests of other factors. *J Anim Sci* 75: 2192–2196.

Driever W, Nüsslein-Volhard C (1989) The bicoid protein is a positive regulator of hunchback transcription in the early *Drosophila* embryo. *Nature* 337: 138–143.

Driscoll SG, Taylor SH (1980) Effects of prenatal maternal estrogen on the male urogenital system. *Obstet Gynecol* 56: 537–542.

Duerr JS, Frisby DL, Gaskin J, Duke A, Asermely K, Huddleston D, Eiden LE, Rand JB (1999) The cat-1 gene of *Caenorhabditis elegans* encodes a vesicular monoamine transporter required for specific monoamine-dependent behaviors. *J Neurosci* 19: 72–84.

Duhon SA, Johnson TE (1995) Movement as an index of vitality: comparing wild type and the age-1 mutant of *Caenorhabditis elegans*. *J Gerontol A Biol Sci Med Sci* 50: B254–261.

Dun RB, Fraser AS (1959) Selection for an invariant character, vibrissae number in the house mouse. *Austral J Biol Sci* 12: 506–523.

Durr D, Spohn H (1998) Brownian motion and microscopic chaos. *Nature* 394: 831–832.

Edelman GM (1987) *Neural Drawinism*, Basic Books, New York.

Edwards IJ, Rudel LL, Terry JG, Kemnitz JW, Weindruch R, Cefalu WT (1998) Caloric restriction in rhesus monkeys reduces low density lipoprotein interaction with arterial proteoglycans. *J Gerontol A Biol Sci Med Sci* 53: B443–B448.

Effros RB (1998) Replicative senescence in the immune system: impact of the Hayflick limit on T-cell function in the elderly. *Am J Hum Genet* 62: 1003–1007.

Eichenlaub-Ritter U, Chandley AC, Gosden RG (1988) The CBA mouse as a model for age-related aneuploid in man: studies of oocyte maturation, spindle formation, and chromosome alignment during meiosis. *Chromosoma* 96: 220–226.

Ellinwood WE, Resko JA (1980) Sex differences in biologically active and immunoreactive gonadotropins in the fetal circulation of rhesus monkeys. *Endocrinology* 107: 902–907.

Ellsworth JL, Schimke RT (1990) On the frequency of metaphase chromosome aberrations in the small intestine of aged rats. *J Gerontol* 45: B94–B100.

Emlen JM, Freeman DC, Graham JH (1993) Non linear growth dynamics and the origins of fluctuating asymmetry. *Genetica* 89: 77–96.

Enver, T, Greaves M (1998) Loops, lineage, and leukemia. *Cell* 94: 9–12.

Eriksson PS, Perfilieva E, Bjork-Eriksson T, Alborn AM, Nordborg C, Peterson DA, Gage FH (1998) Neurogenesis in the adult human hippocampus. *Nature Med* 4: 1313–1317.

Evan G, Littlewood T (1998) A matter of life and cell death. *Science* 281: 1317–1322.

Even MD, Dhar MG, vom Saal FS (1992) Transport of steroids between fetuses via amniotic fluid in relation to the intrauterine position phenomenon in rats. *J Reprod Fertil* 96: 709–716.

Even MD, Laughlin MH, Krause GF, vom Saal FS (1994) Differences in blood flow to uterine segments and placentae in relation to sex, intrauterine location and side in pregnant rats. *J Reprod Fertil* 102: 245–252.

Faber KA, Ayyash L, Dixon S, Hughes CL Jr (1993) Effect of neonatal diethylstilbestrol exposure on volume of the sexually dimorphic nucleus of the preoptic area of the hypothalamus and pituitary responsiveness to gonadotropin-releasing hormone in female rats of known anogenital distance at birth. *Biol Reprod* 48: 947–951.

Fabricant JD, Parkening TA (1978) Sperm morphology and cytogenic studies in aging C57BL/6 mice. *Mech Age Dev* 66: 485–489.

Faddy MJ, Gosden RG (1995) A mathematical model of follicle dynamics in the human ovary. *Hum Reprod* 10: 770–775.

Faddy MJ, Gosden RG (1996) A model conforming the decline in follicle numbers to the age of menopause in women. *Hum Reprod* 11: 1484–1486.

Faddy MJ, Gosden RG, Edwards RG (1983) Ovarian follicle dynamics in mice: a comparative study of three inbred strains and an F1 hybrid. *J Endocrinol* 96: 23–33.

Failla G (1958) The aging process and carcinogenesis. *Ann NY Acad Sci* 71: 1124–1135.

Falconer DS, Mackay TFC (1996) *Introduction to Quantitative Genetics*, 4th ed., Longman, London.

Fallin D, Gauntlett AC, Scibelli P, Cai X, Duara R, Gold M, Crawford F, Mullan M (1997) No association between the very low density lipoprotein receptor gene and late-onset Alzheimer's disease nor interaction with the apolipoprotein E gene in population-based and clinic samples. *Genet Epidemiol* 14: 299–305.

Ferrari S, Rizzoli R, Bonjour JP (1998a) Heritable and nutritional influences on bone mineral mass. *Aging Clin Exp Res* 10: 205–213.

Ferrari S, Rizzoli R, Slosman D, Bonjour JP (1998b) Familial resemblance for bone mineral mass is expressed before puberty. *J Clin Endocrinol Metab* 83: 358–361.

Festing MFW (1996) Origins and characteristics of inbred strains of mice. In: *Genetic Variants and Strains of the Laboratory Mouse* (ed. Lyon MF, Rastan S, Brown ADM), pp. 1537–1596, Oxford University Press, New York.

Finch CE (1980) The relationships of aging changes in the basal ganglia to manifestations of Huntington's chorea. *Ann Neurol* 7: 406–411.

Finch CE (1982) Rodent models for aging processes in the human brain. In: *Aging, Vol. 19: Alzheimer's Disease: A Report of Progress* (ed. Corkin S et al.), pp. 249–257, Raven Press, New York.

Finch CE (1990) *Longevity, Senescence, and the Genome*. Chicago University Press, Chicago.

Finch CE (1993) Neuron atrophy during aging: programmed or sporadic? *Trends Neurosci* 16: 104–110. Corrigenda: aged dogs, space, and scissors. *ibid* 16: 352.

Finch CE (1996) Biological bases for plasticity during aging of individual life histories. In: *The Life Span Development of Individuals: Biological and Psychosocial Perspectives, a Synthesis* (ed. Magnusson D), pp 488–512, Cambridge University Press, Cambridge.

Finch CE (1997) Longevity: is everything under control? Non-genetic and non-environmental sources of variation. In: *Longevity: To The Limits and Beyond* (ed. Robine J-M, Vaupel J, Jeune B, Allard M), pp 165–178, Springer-Verlag, Heidelburg.

Finch CE (1998) Variations in senescence and longevity include the possibility of negligible senescence. *J Gerontol* 53A: B235–B239.

Finch CE, Goodman MF (1997) Relevance of "adaptive" mutations arising in non-dividing cells in microorganisms to age-related changes in mutant phenotypes of neuron. *Trends Neurosci* 20: 501–507.

Finch CE, Landfield PW (1985) Neuroendocrine and autonomic function in aging mammals. In: *Handbook of the Biology of Aging, 2nd ed.* (ed. Finch CE, Schneider EL), pp 79–90, Van Nostrand, New York.

Finch CE, Loehlin JC (1998) Environmental influences that may precede fertilization: a first examination of the prezygotic hypothesis from maternal age influences on twins. *Behav Genet* 28: 101–106.

Finch CE, McNeill T (1993) Neuroplasticity of the hippocampus and striatum: for aging and neurodegenerative disease. In: *Synaptic Plasticity: Molecular, Cellular, and Functional Aspects* (ed. Davis J, Baudry M, Thompson R), pp 45–72, MIT Press, Cambridge, Mass.

Finch CE, Morgan TE (1997) Food restriction and brain aging. *Adv Cell Aging Gerontol* 2: 279–297.

Finch CE, Pike MC (1996) Maximum lifespan predictions from the Gompertz mortality model. *J Gerontol* 51: B183–B194.

Finch CE, Rose MR (1995) Hormones and the physiological architecture of life history evolution. *Q Rev Biol* 70: 1–52.

Finch CE, Sapolsky RM (1999) The reproductive schedule, ApoE, and the evolution of Alzheimer disease. *Neurobiol Aging*, in press.

Finch CE, Tanzi RE (1997) The genetics of aging. *Science* 278: 407–411.

Finch CE, Felicio LS, Flurkey K, Gee DM, Mobbs C, Nelson JF, Osterburg HH (1980) Studies on ovarian-hypothalamic-pituitary interactions during reproductive aging in C57BL/6J mice. In: *Brain–Endocrine Interaction, IV: Neuropeptides in Development and Aging* (ed. Scott D, Sladek JL Jr), *Peptides* (vol 1, suppl 1), pp 163–175, ANKHO International, Fayetteville, NY.

Finch CE, Felicio LS, Mobbs CV, Nelson JF (1984) Ovarian and steroidal influences on neuroendocrine aging processes in female rodents. *Endocr Rev* 5: 467–497.

Finch CE, Pike MC, Witten M (1990) Slow mortality rate accelerations during aging in animals approximate that of humans. *Science* 249: 902–905.

Flaws JA, Abbud R, Mann RJ, Nilson JH, Hirshfield AN (1997) Chronically elevated luteinizing hormone depletes primordial follicles in the mouse ovary. *Biol Reprod* 57: 1233–1237.

Forget H, Cohen H (1994) Life after birth: the influence of steroid hormones on cerebral structure and function is not fixed prenatally. *Brain Cognit* 26: 243–248.

Forsén T, Eriksson JG, Tuomilehto J, Teramo K, Osmond C, Barker DJP (1997) Mother's weight in pregnancy and coronary heart disease in a cohort of Finnish men: follow up study. *Br Med J* 315: 837–840.

Fraser S, Keynes R, Lumsden A (1990) Segmentation in the chick embryo hindbrain is defined by cell lineage restrictions. *Nature* 344: 431–435.

Fredholm M, Policastro PF, Wilson MC (1991) The dispersion of defective endogenous murine retroviral elements suggests retrotransposition-mediated amplification. *DNA Cell Biol* 10: 713–722.

Freeman ME (1994) In: *Physiology of Reproduction*. Knobil E et al., eds, 2nd ed., Chapter 46, vol 2, pp 613–659, Raven Press, New York.

Furlow FB, Armijo-Prewitt T, Gangestad SW, Thornhill R (1997) Fluctuating asymmetry and psychometric intelligence. *Proc Soc R Soc Lond [Biol]* 264: 823–829.

Gage FH, Kempermann G, Palmer TD, Peterson DA, Ray J (1998) Multipotent progenitor cells in the adult dentate gyrus. *J Neurobiol* 36: 249–266.

Gale RE, Fielding AK, Harrison CN, Linch DC (1997) Acquired skewing of X-chromosome inactivation patterns in myeloid cells of the elderly suggests stochastic clonal loss with age. *Br J Haematol* 98: 512–519.

Gallagher M, Landfield PW, McEwen B, Meaney MJ, Rapp PR, Sapolsky R, West MJ (1996) Hippocampal neurodegeneration in aging. *Science* 274: 484–485.

Garcia-Castro MI, Anderson R, Heasman J, Wylie C (1997) Interactions between germ cells and extracellular matrix glycoproteins during migration and gonad assembly in the mouse embryo. *J Cell Biol* 138: 471–480.

Garkavtsev I, Hull C, Riabowol K (1998) Molecular aspects of the relationship between cancer and aging: tumor suppressor activity during cellular senescence. *Exp Gerontol* 33: 81–94.

Garrington TP, Johnson GL (1999) Organization and regulation of mitogen-activated protein kinase signaling pathways. *Curr Opinion Cell Biol* 11: 211–218.

Gärtner K (1990) A third component causing random variability beside environment and genotype. A reason for the limited success of a 30 year long effort to standardize laboratory animals? *Lab Anim* 24: 71–77.

Gärtner K, Baunack E (1981) Is the similarity of monozygotic twins due to genetic factors alone. *Nature* 292: 646–647.

Gaspard P, Briggs ME, Francis MK, Sengers JV, Gammon RW, Dorfman JR, Calabrese RV (1998) Experimental evidence for microscopic chaos. *Nature* 394: 865–868.

Gatz M, Pedersen NL, Berg S, Johannson B, Johannson K, Mortimer JA, Posner SF, Vitanen M, Winblad B, Ahlbom A (1997) Heritability for Alzheimer's disease: the study of dementia in Swedish twins. *J Gerontol Med Sci* 52A: M117–M125.

Gazit Y, Berk DA, Leunig M, Baxter LT, Jain RK (1995) Scale-invariant behavior and vascular network formation in normal and tumor tissue. *Phys Rev Letters* 75: 2428–2431.

Geissmann T (1990) Familial incidence of multiple births in a colony of chimpanzees (*Pan troglodytes*). *J Med Primatol* 19: 467–478.

Gelman R, Watson A, Bronson R, Yunis E (1988) Murine chromosomal regions correlated with longevity. *Genetics* 118: 693–704.

Gems D, Sutton AJ, Sundermeyer ML, Albert PS, King KV, Edgley ML, Larsen PL, Riddle DL (1998) Two pleiotropic classes of daf-2 mutation affect larval arrest, adult behavior, reproduction and longevity in *Caenorhabditis elegans*. *Genetics* 150: 129–155.

George FW, Wilson JD (1994) Sex determination and differentiation. In: *The Physiology of Reproduction*, 2nd ed. (ed. Knobil E, Neill JD), vol 1, pp 3–28, Raven Press, New York.

Gerich JE (1998) The genetic basis of type 2 diabetes mellitus: impaired insulin secretion versus impaired insulin sensitivity. *Endocr Rev* 19: 491–503.

Geschwind N, Galaburda AM (1985a) Biological mechanisms, associations, and pathology: I. A hypothesis and a program for research. *Arch Neurol* 42: 428–459.

Geschwind N, Galaburda AM (1985b) Biological mechanisms, associations, and pathology: II. A hypothesis and a program for research. *Arch Neurol* 42: 521–552.

Geschwind N, Galaburda AM (1985c) Biological mechanisms, associations, and pathology. III. A hypothesis and a program for research. *Arch Neurol* 42: 634–654.

Gho M, Lecourtois M, Géraud G, Posakony JW, Schweisguth F (1996) Subcellular localization of suppressor of *Hairless* in *Drosophila* sense organ cells during Notch signalling. *Development* 122: 1673–1682.

Gilissen E, Zilles K (1996) The calcarine sulcus as an estimate of the total volume of human striate cortex: a morphometric study of reliability and intersubject variability. *J Hirnforsch* 37: 57–66.

Gilissen E, Iba-Zizen MT, Stievenart JL, Lopez A, Trad M, Cabanis EA, Zilles K (1995) Is the length of the calcarine sulcus associated with the size of the human visual cortex? A morphometric study with magnetic resonance tomography. *J Hirnforsch* 36: 451–459.

Gill WB, Schumacher GF, Bibbo M (1976) Structural and functional abnormalities in the sex organs of male offspring of mothers treated with diethylstilbestrol (DES). *J Reprod Med* 16: 147–153.

Gill WB, Schumacher GF, Bibbo M, Straus FH 2d, Schoenberg HW (1979) Association of diethylstilbestrol exposure in utero with cryptorchidism, testicular hypoplasia and semen abnormalities. *J Urol* 122: 36–39.

Gillespie DT (1977) Exact stochastic simulation of coupled chemical reactions. *J Phys Chem* 81: 2340–2361.

Gillespie DT (1992) A rigorous derivation of the chemical master equation. *Physica A* 188: 404–445.

Gilmore JH, Perkins DO, Kliewer MA, Hage ML, Silva SG, Chescheir NC, Hertzberg BS, Sears CA (1996) Fetal brain development of twins assessed in utero by ultrasound: implications for schizophrenia. *Schizophr Res* 19: 141–149.

Godfrey K, Robinson S, Barker DJP, Osmond C, Cox V (1996) Maternal nutrition in early and late pregnancy in relation to placental and fetal growth. *Br Med J* 312: 410–414.

Goedert M, Crowther A, Spillantini MG (1998) Tau mutations cause frontotemporal dementias. *Neuron* 21: 955–958.

Goodman CS (1976) Constancy and uniqueness in a large population of small inter-neurons. *Science* 193: 502–504.

Goodman CS (1978) Isogenic grass-hoppers: genetic variability in the morphology of identified neurons. *J Comp Neurol* 182: 681–706.

Goodman MF (1997) Hydrogen bonding revisited: geometric selection as a principal determinant of DNA replication fidelity. *Proc Natl Acad Sci USA* 94: 10493–10495.

Goodman MF, Creighton S, Bloom LB, Petruska J (1993) Biochemical basis of DNA replication fidelity. *Crit Rev Biochem Mol Biol* 28: 83–126.

Gorski RA (1996) Gonadal hormones and the organization of brain structure and function. In: *The Life Span Development of Individuals: Biological and Psychosocial Perspectives, a Synthesis* (ed. Magnusson D), pp 315–340, Cambridge University Press, Cambridge.

Gosden RG (1985) *The Biology of Menopause: The Cause and Consequence of Ovarian Aging*, Academic Press, San Diego.

Gosden RG, Faddy MJ (1998) Biological bases of premature ovarian failure. *Reprod Fertil Dev* 10: 73–78.

Gosden RG, Laing SC, Felicio LS, Nelson JF, Finch CE (1983) Imminent oocyte exhaustion and reduced follicular recruitment mark the transition to acyclicity in aging C57BL/6J mice. *Biol Reprod* 28: 255–260.

Gosden RG, Wade JC, Fraser HM, Sandow J, Faddy MJ (1997) Impact of congenital or experimental hypogonadotrophism on the radiation sensitivity of the mouse ovary. *Hum Reprod* 12: 2483–2488.

Gossen JA, de Leeuw WJF, Bakker AQ, Vijg J (1993) DNA sequence analysis of spontaneous mutations at a lacZ transgene integrated on the mouse X-chromosome. *Mutagenesis* 8: 243–247.

Gossen JA, Martus HJ, Wei JY, Vijg J (1995) Spontaneous and x-ray–induced deletion mutations in a lacZ plasmid–based transgenic mouse model. *Mutat Res* 331: 89–97.

Götz M, Williams BP, Bolz J, Price J (1995) The specification of neuronal fate: a common precursor for the neurotransmitter subtypes in the rat cerebral cortex in vitro. *Eur J Neurosci* 7: 889–898.

Goy RW, Bercovitch FB, McBrair MC (1988) Behavioral masculinization is independent of genital rhesus macaques. *Hormones Behav* 22: 552–571.

Grammar K, Thornhill R (1994) Human facial attractiveness and sexual selection: the role of symmetry and averageness. *J Comp Psychol* 108: 233–242.

Gray MD, Shen JC, Kamath-Loeb AS, Blank A, Sopher BL, Martin GM, Oshima J, Loeb LA (1997) The Werner syndrome protein is a DNA helicase. *Nat Genet* 17: 100–103.

Greco TL, Payne AH (1994) Ontogeny of expression of the genes for steroidogenic enzymes P450 side-chain cleavage, 3α-hydroxysteroid dehydrogenase, P450 17α-hydroxylase/C17-20 lyase, and P450 aromatase in fetal mouse gonads. *Endocrinology* 135: 262–268.

Gridley T (1997) Notch signaling in vertebrate development and disease. *Mol Cell Neurosci* 9: 103–108.

Grove EA, Kirkwood TBL, Price J (1992) Neuronal precursor cells in the rat hippocampal formation contribute to more than one cytoarchitectonic area. *Neuron* 8: 217–229.

Hales CN (1997) Fetal and infant growth and impaired glucose tolerance in adulthood: the "thrifty phenotype" hypothesis revisited. *Acta Paediatr [Suppl]* 422: 73–77.

Hales CN, Desai M, Ozanne SF, Crowther NJ (1996) Fishing in the stream of diabetes: from measuring insulin to the control of fetal organogenesis. *Biochem Soc Trans* 24: 341–350.

Hamdy S, Aziz Q, Rothwell JC, Singh KD, Barlow J, Hughes DG, Tallis RC, Thompson DG (1996) The cortical topography of human swallowing musculature in health and disease. *Nat Med* 2: 1217–1224.

Hamilton WD (1996) The moulding of senescence by natural selection. *J Theor Biol* 12: 12–45.

Hampson E, Moffat SD (1994) Is testosterone related to spatial cognition and hand preference in humans? *Brain Cognit* 26: 255–266.

Haney AF, Newbold RR, Fetter BF, McLachlan JA (1986) Paraovarian cysts associated with prenatal diethylstilbestrol exposure. Comparison of the human with a mouse model. *Am J Pathol* 124: 405–411.

Hanselaar A, van Loosbroek M, Schuurbiers O, Helmerhorst T, Bulten J, Bernhelm J (1997) Clear cell adenocarcinoma of the vagina and cervix. An unpdate of the central Netherlands registry showing twin age incidence peaks. *Cancer* 79: 2229–2236.

Haqq CM, Donahow PK (1998) Regulation of sexual dimorphism in mammals. *Physiol Rev* 78: 1–33.

Hardy ICW (1995) Protagonists of polyembryony. *Trends Res Ecol Evol* 10: 179–180.

Hardy J (1997) The Alzheimer family of diseases: many etiologies, one pathogenesis? *Proc Natl Acad Sci USA* 94: 2095–2097.

Harley CB (1997) Human ageing and telomeres. *Ciba Found Symp* 211: 129–144.

Harley CB, Sherwood SW (1997) Telomerase, checkpoints and cancer. *Cancer Surv* 29: 263–284.

Harman SM, Talbert GB (1985) Reproductive aging. In: *Handbook of the Biology of Aging, 2nd ed.* (ed. Finch CE, Schneider EL), pp 457–510, Von Nostrand, New York.

Harrington MG, Coffman JA, Calzone FJ, Hood LE, Britten RJ, Davidson EH (1992) Complexity of sea urchin embryo nuclear proteins that contain basic domains. *Proc Natl Acad Sci USA* 89: 6252–6256.

Harrison DE (1979) Mouse erythropoietic stem cell lines function normally 100 months: loss related to number of transplantations. *Mech Ageing Dev* 9: 427–433.

Harrison DE (1984) Do hemopoietic cells age? *Monogr Dev Biol* 17: 21–41.

Harrison DE, Archer JR (1989) Natural selection for extended longevity from food restriction. *Growth Dev Aging* 53: 3–6.

Harrison DE, Astle CM, DeLaittre JA (1978) Loss of proliferative capacity in immunohemopoietic cells caused by serial transplantation rather than aging. *J Exp Med* 147: 1526–1531.

Harrison DE, Astle CM, Stone M (1989) Numbers and functions of transplantable primitive immunopoietic stem cells. Effects of age. *J Immunol* 142: 3833–3840.

Hartenstein V, Posakony JW (1989) Development of adult sensilla on the wing and notum of *Drosophila melanogaster*. *Development* 107: 389–405.

Hartley D (1996) *Drosophila* inherit diseases. *Nat Genet* 13: 133–134.

Harvey AW (1998) Genes for asymmetry easily overruled. *Nature* 392: 345–346.

Haskell SG, Richardson ED, Horwitz RI (1997) The effect of estrogen replacement therapy on cognitive function in women: a critical review of the literature. *J Clin Epidemiol* 50: 1249–1264.

Hayflick L, Moorhead PS (1961) The serial cultivation of human diploid cell strains. *Exp Cell Res* 25: 585–621.

Headley JA, Theriault RL, LeBlanc AD, Vassilopoulou-Sellin R, Hortobagyi GN (1998) Pilot study of bone mineral density in breast cancer patients treated with adjuvant chemotherapy. *Cancer Invest* 16: 6–11.

Heitzler P, Simpson P (1991) The choice of cell fate in the epidermis of *Drosophila*. *Cell* 64: 1083–1092.

Helbock HJ, Beckman KB, Shigenaga MK, Walter PB, Woodall AA, Yeo HC, Ames BN (1998) DNA oxidation matters: the HPLC-electrochemical detection assay of 8-oxo-deoxyguanosine and 8-oxo-guanine. *Proc Natl Acad Sci USA* 95: 288–293.

Heller DA, Pedersen NL, deFaire U, McClearn GE (1994) Genetic and environmental correlations among serum lipids and apolipoproteins in elderly twins reared together and apart. *Am J Hum Genet* 55: 1255–1267.

Hellwig B, Schüz A, Aertsen A (1994) Synapses on axon collaterals of pyramidal cells are spaced at random intervals: a Golgi study in the mouse cerebral cortex. *Biol Cybern* 71: 1–12.

Henderson BA, Berenbaum SA (1997) Sex-typed play in opposite-sex twins. *Devel Psychobiol* 31: 115–123.

Henderson BE, Ross RK, Pike MC (1993) Hormonal chemoprevention of cancer in women. *Science* 259: 653–638.

Henderson SA, Edwards RG (1968) Chiasma frequency and maternal age in mammals. *Nature* 218: 22–28.

Hendry JH, Roberts SA, Potten CS (1992) The clonogen content of murine intestinal crypts: dependence on radiation dose used in its determination. *Radiat Res* 132: 115–119.

Herbst AL, Bern HA (eds.) (1981) *Developmental Effects of Diethylstilbestrol (DES) in Pregnancy*, Thieme-Stratton, New York.

Herbst AL, Ulfelder H, Poskanzer DC (1971) Adenocarcinoma of the vagina. Association of maternal stibestrol therapy with tumor appearance in young women. *N Engl J Med* 284: 878–881.

Herriot R, Walker F (1989) Age-related disposition of amyloid P component in normal human testis. *J Pathol* 157: 11–14.

Herrnstein R, Murray C (1994) *The Bell Curve*, Free Press, New York.

Herskind AM, McGue M, Holm NV, Sorensen TI, Harvald B, Vaupel JW (1996) The heritability of human longevity: a population-based study of 2872 Danish twin pairs born 1870–1900. *Hum Genet* 197: 319–323.

Higuchi K (1997) Genetic characterization of senescence-accelerated mouse (SAM). *Exp Gerontol* 32: 129–138.

Hirshfield AN (1988) Size-frequency analysis of atresia in cycling rats. *Biol Reprod* 38: 1181–1188.

Hirshfield AN (1991) Development of follicles in the mammalian ovary. *Int Rev Cytol* 124: 43–101.

Hirshfield AN (1992) Heterogeneity of cell populations that contribute to the formation of primordial follicles in rats. *Biol Reprod* 47: 466–472.

Hirshfield AN (1994) Relationship between the supply of primordial follicles and the onset of follicular growth in rats. *Biol Reprod* 50: 421–428.

Hirshfield AN (1997) Overview of ovarian follicular development: considerations for the toxicologist. *Environ Mol Mutagen* 29: 10–15.

Hirshfield AN, DeSanti AM (1995) Patterns of ovarian cell proliferation in rats during the embryonic period and the first three weeks postpartum. *Biol Reprod* 53: 1201–1221.

Hirvonen J, Yla-Herttuala S, Laaksonen H, Mottonen M, Nikkari T, Pesonen E, Raekallio J, Akerblom HK (1985) Coronary intimal thickenings and lipids in Finnish children who died violently. *Acta Paediatr Scand [Suppl]* 318: 221–224.

Holinka CF, Finch CE (1978) Dextral bias in the induced decidual response after ovariectomy and in implantation sites in the C57BL/6J mouse uterus. *Biol Reprod* 18: 418–420.

Holinka CF, Tseng Y-C, Finch CE (1979a) Reproductive aging in C57BL/6J mice: Plasma progesterone, viable embryos and resorption frequency throughout pregnancy. *Biol Reprod* 20: 1201–1211.

Holinka CF, Tseng Y-C, Finch CE (1979b) Impaired preparturitional rise of plasma estradiol in aging C57BL/6J mice. *Biol Reprod* 21: 1009–1013.

Holliday R (1986) *Genes, Proteins and Cellular Aging*. Van Nostrand Reinhold, New York.

Holliday R (1987) The inheritance of epigenetic defects. *Science* 238: 163–170.

Holliday R (1989) Food, reproduction and longevity: is the extended lifespan of calorie restricted animals an evolutionary adaptation? *BioEssays* 10: 125–127.

Holliday R, Huschtscha LI, Tarrant GM, Kirkwood TBL (1997) Testing the commitment theory of cellular ageing. *Science* 198: 366–372.

Holloway DM, Harrison LG (1999) Suppression of positional errors in biological development. *Math. Biosci.* 156: 271–290.

Hollyday M, Hamburger V (1976) Reduction of the naturally occurring motor neuron loss by enlargement of the periphery. *J Comp Neurol* 170: 311–320.

Holt PR, Yeh K (1989) Small intestinal crypt cell proliferation rates are increased in senescent rats. *J Gerontol* 47: B9–B14.

Holt PR, Yeh K, Kotler DP (1988) Altered control of proliferation in proximal small intestine. *Proc Natl Acad Sci USA* 85: 2771–2775.

Hong M, Zhukareva V, Vogelsberg-Ragaglia V, Wszolek Z, Reed L, Miller BI, Geschwind DH, Bird TD, McKeel D, Goate A, Morris JC, Wilhelmsen KC, Schellenberg GD, Trojanowski JQ, Lee VM (1998) Mutation-specific functional impairments in distinct tau isoforms of hereditary FTDP-17. *Science* 282: 1914–1917.

Hong Y, de Faire U, Heller DA, McClearn GE, Pedersen NL (1994) Genetic and environmental influences on blood pressure in elderly twins. *Hypertension* 24: 663–670.

Hook EB (1981) Rates of chromosome abnormalities at different maternal ages. *Obstet Gynecol* 58: 282–285.

Hopfield JJ (1974) Kinetic proofreading: a new mechanism for reducing errors in biosynthetic processes requiring high specificity. *Proc Natl Acad Sci USA* 71: 4135–4139.

Hopper JL, Seeman E (1994) The bone density of female twins discordant for tobacco use. *N Engl J Med* 330: 387–392.

Houlden H, Crook R, Backhovens H, Prihar G, Baker M, Hutton M, Rossor M, Martin JJ, Van Broeckhoven C, Hardy J (1998) ApoE genotype is a risk factor in nonpresenilin early-onset Alzheimer's disease families. *Am J Med Genet* 81: 117–121.

Houlé D (1992) Comparing evolvability and variability of quantitative traits. *Genetics* 130: 195–204.

Hoyer AP, Grandjean P, Jorgensen T, Brock JW, Hartvig HB (1998) Organochlorine exposure and risk of breast cancer. *Lancet* 352: 1816–1820.

Hrubec Z, Robinette CD (1984) The study of human twins in medical research. *N Engl J Med* 310: 435–441.

Hu Q, Kukull WA, Bressler SL, Gray MD, Cam JA, Larson EB, Martin GM, Deeb SS (1998) The human FE65 gene: genomic structure and an intronic biallelic polymorphism associated with sporadic dementia of the Alzheimer type. *Human Genet* 103: 295–303.

Hustler JJ, Loftus WC, Gazzaniga MS (1998) Individual variation of cortical surface area asymmetries. *Cereb Cortex* 8: 11–17.

Hyatt BA, Lohr JL, Yost HJ (1996) Initiation of vertebrate left–right axis formation by maternal Vg1. *Nature* 384: 62–65.

Iguchi T, Fukazawa Y, Bern HA (1995) Effects of sex hormones on oncogene expression in the vagina and on development of sexual dimorphism of the pelvis and anococcygeus muscle in the mouse. *Environ Health Perspect* 103 [Suppl 7]: 79–82.

Ishii N, Fujii M, Hartman PS, Tsuda M, Yasuda K, Senoo-Matsuda N, Yanase S, Ayusawa D, Suzuki K (1998) A mutation in succinate dehydrogenase cytochrome b causes oxidative stress and ageing in nematodes. *Nature* 394: 694–697.

Ito M, Lang TF, Jergas M, Ohki M, Takada M, Nakamura T, Hayashi K, Genant HK (1997) Spinal trabecular bone loss and fracture in American and Japanese women. *Calcif Tissue Int* 61: 123–128.

Ivanova R, Henon N, Lepage V, Charron D, Vicaut E, Schachter F (1998) HLA-DR alleles display sex-dependent effects on survival and discriminate between individual and familial longevity. *Hum Mol Genet* 7: 187–194.

Jacobson M (1991) *Developmental Neurobiology*, 3rd ed., Plenum Press, New York.

Jakobsen H, Torp-Pedersen S, Juul N (1988) Ultrasonic evaluation of age-related human prostatic growth and development of benign prostatic hyperplasia. *Scand J Urol Nephrol [Suppl]* 107: 26–31.

Jan YN, Jan LY (1998) Asymmetric cell division. *Nature* 392: 775–778.

Jankowski RA, Ilstad ST (1997) Chimerism and tolerance: from freemartin cattle and neonatal mice to humans. *Human Immunol* 52: 155–161.

Jazwinski SM (1996) Longevity, genes, and aging. *Science* 273: 54–59.

Jenner P (1998) Oxidative mechanisms in nigral cell death in Parkinson's disease. *Mov Disord* 13 [Suppl 1]: 24–34.

Jennings B, de Celis J, Delidakis C, Preiss A, Bray S (1995) Role of *Notch* and *achaete-scute* complex in the expression of *Enhancer of split* bHLH proteins. *Development* 121: 3745–3752.

Jiménez R, Burgos M (1998) Mammalian sex determination: joining the pieces of the genetic puzzle. *BioEssays* 20: 696–699.

Johnson TE (1987) Aging can be genetically dissected into component processes using long-lived lines of *Caenorhabditis elegans*. *Proc Natl Acad Sci USA* 84: 3777–3781.

Johnson TE, Wood WB (1982) Genetic analysis of life-span in *Caenorhabditis elegans*. *Proc Natl Acad Sci USA* 79: 6603–6607.

Jokela P, Portin P (1991) Effect of extra Y chromosome on number and fluctuating asymmetry of sternopleural bristles in *Drosophila melanogaster*. *Heridatas* 114: 177–187.

Jones D, Gonzalez-Lima F, Crews D, Galef BG Jr, Clark MM (1997) Effects of intrauterine position on the metabolic capacity of the hypothalamus of female gerbils. *Physiol Behav* 61: 513–519.

Jones EC, Krohn PL (1961a) The relationships between age, numbers of oocytes, and fertility in virgin and multiparous mice. *J Endocrinol* 21: 469–496.

Jones EC, Krohn PL (1961b) The effect of hypophysectomy on age changes in the ovaries of mice. *J Endocrinol* 21: 497–508.

Jones PH (1997) Epithelial stem cells. *BioEssays* 19: 683–690.

Jost A (1965) Gonadal hormones in the sex differentiation of the mammalian fetus. In: *Organogenesis* (ed. DeHaan R, Ursprung H), pp 611–628, Holt, Rinehart, Winston, New York.

Jouandet ML, Tramp MJ, Herron DM, Hermann A, Loftus WC, Bazell J, Gazzaniga MS (1989) Brain prints: computer generated cerebral cortex in vivo. *J Cogn Neurosci* 1: 88–117.

Juniewicz PE, Berry SJ, Coffey DS, Strandberg JD, Ewing LL (1994) The requirement of the testis in establishing the sensitivity of the canine prostate to develop benign prostatic hyperplasia. *J Urol* 152: 996–1001.

Kaas JH (1991) Plasticity of sensory motor maps in adult mammals. *Annu Rev Neurosci* 14: 137–167.

Kahn BB, Rossetti L (1998) Type 2 diabetes—who is conducting the orchestra? *Nature Genet* 20: 223–225.

Kahn HS, Narayan VMK, Valdez R (1998) Prenatal exposure to famine and health in later life. *Lancet* 351: 1360–1362.

Kaiser M, Gasser M, Ackermann R, Stearns SC (1997) P-element inserts in transgenic flies: a cautionary tale. *Heredity* 78: 1–11.

Kandel ER, Schwart JH, Jessell TM (1992) *Principles of Neural Science*, 3rd ed., Appleton and Lange.

Kang DH, Davidson RJ, Coe CL, Wheeler RE, Tomarken AJ, Ershler WB (1991) Frontal brain asymmetry and immune function. *Behav Neurosci* 105: 860–869.

Kannisto V, Christensen K, Vaupel JW (1997) No increased mortality in later life for cohorts born during famine. *Am J Epidemiol* 145: 987–994.

Kapahi P, Boulton ME, Kirkwood TBL (1999) Enhanced resistance to multiple cellular stresses in long-lived mammalian species. *Free Radical Biol Med* 26: 495–500.

Kardia SLR, Stengård J, Templeton A (1999) An evolutionary perspective on the genetic architecture of susceptibility to cardiovascular disease. In: *Evolution in Health and Disease* (ed. Stearns SC), pp 231–245, Oxford University Press, New York.

Kasuga A, Falorni A, Maruyama T, Ozawa Y, Grubin CE, Matsubara K, Takei I, Saruta T, Scheynius A, Lernmark A (1996) HLA class II is associated with the frequency of glutamic acid decarboxylase M(r) 65,000 autoantibodies in Japanese patients with insulin-dependent diabetes mellitus. *Acta Diabetol* 33: 108–113.

Kaur P, Potten CS (1986) Cell migration velocities in the crypts of the small intestine after cytotoxic insult are not dependent on mitotic activity. *Cell Tiss Kinet* 19: 601–610.

Kawas C, Resnick S, Morrison A, Brookmeyer R, Corrada M, Zonderman A, Bacal C, Lingle DD, Metter E (1997) A prospective study of estrogen replacement therapy and the risk of developing Alzheimer's disease: the Baltimore Longitudinal Study of Aging. *Neurology* 48: 1517–1521. [published erratum appears in *Neurology* (1998) 51: 654.]

Keet MP, Jaroszewicz AM, Lombard CJ (1986) Follow-up study of physical growth of monozygous twins with discordant within-pair birth weights. *Pediatrics* 77: 336–344.

Kell EN, Shaw AS, Allen PM (1998) Fidelity of T cell activation through multi-step T cell receptor ζ phosphorylation. *Science* 281: 572–574.

Khan AM, Walker F (1984) Age related detection of tissue amyloid P in the skin. *J Pathol* 143: 183–186.

Kieser JA (1992) Fluctuating odontometric asymmetry and maternal alcohol consumption. *Ann Human Biol* 19: 513–520.

Kilner R (1998) Primary and secondary sex ratio manipulation by zebra finches. *Anim Behav* 56: 155–164.

Kilpatrick TJ, Bartlett PF (1993) Cloning and growth of multipotential neural precursors: requirements for proliferation and differentiation. *Neuron* 10: 255–265.

Kimble J, Ward S (1988) Germ-line development and fertilization. In: *The Nematode* Caenorhabditis elegans (ed. Wood WB), pp 191–213, Cold Spring Harbor Laboratory Press, Cold Spring, NY.

Kimura H, Suzuki Y, Gejyo F, Karasawa R, Miyazaki R, Suzuki S, Arakawa M (1998) Apolipoprotein E4 reduces risk of diabetic nephropathy in patients with NIDDM. *Am J Kidney Dis* 31: 666–673.

King CM, Gillespie ES, McKenna PG, Barnett YA (1994) An investigation of mutation as a function of age in humans. *Mutat Res* 316: 79–90.

King CR, Magenis E, Bennett S (1978) Pregnancy and the Turner syndrome. *Obstet Gynecol* 52: 617–624.

King RA, Rotter JI, Motulsky AG (1992) *The Genetic Basis of Common Diseases*, Oxford University Press, New York.

Kipling D (1995) *The Telomere*. Oxford University Press, Oxford.

Kipling D (1997) Telomere structure and telomerase expression during mouse development and tumorigenesis. *Eur J Cancer* 33: 792–800.

Kipling D, Wilson HE, Thomson EJ, Lee M, Perry J, Palmer S, Ashworth A, Cooke HJ (1996) Structural variation of the pseudoautosomal region between and within inbred mouse strains. *Proc Natl Acad Sci USA* 93: 171–175.

Kirkwood TBL (1977) Evolution of ageing. *Nature* 270: 301–304.

Kirkwood TBL (1981) Repair and its evolution: survival versus reproduction. In: *Physiological Ecology, an Evolutionary Approach to Resource Use* (ed. Townsend CR, Calow P), pp 165–189, Blackwell Scientific, Oxford.

Kirkwood TBL (1989) DNA, mutations and aging. *Mutat Res* 219: 1–7.

Kirkwood TBL (1992) Comparative lifespans of species: why do species have the life spans they do? *Am J Clin Nutrit* 55: S1191–S1195.

Kirkwood TBL (1999) Evolution, molecular biology, and mortality plateaus. In: *Alfred Benzon Symposium, Molecular Biology of Aging* (ed. Bohr VA, Clark BFC, Stevnsner T) pp. 383–390, Munksgard, Copenhagen.

Kirkwood TBL, Cremer T (1982) Cytogerontology since 1881: a reappraisal of August Weismann and a review of modern progress. *Hum Genet* 60: 101–121.

Kirkwood TBL, Franceschi C (1992) Is ageing as complex as it would appear? New perspectives in ageing research. *Ann NY Acad Sci* 663: 412–417.

Kirkwood TBL, Holliday R (1975) Commitment to senescence: a model for the finite and infinite growth of diploid and transformed human fibroblasts in culture. *J Theor Biol* 53: 481–496.

Kirkwood TBL, Holliday R (1978) A stochastic model for the commitment of human cells to senescence. In: *Biomathematics and Cell Kinetics* (ed. Valleron AJ, Macdonald PDM), pp 161–172, Elsevier/North-Holland, Amsterdam.

Kirkwood TBL, Holliday R (1979) The evolution of ageing and longevity. *Proc R Soc Lond [Biol]* 205: 531–546.

Kirkwood TBL, Rose MR (1991) Evolution of senescence: late survival sacrificed for reproduction. *Philos Trans R Soc Lond [Biol]* 332: 15–24.

Kirkwood TBL, Holliday R, Rosenberger RF (1984) Stability of the cellular translation process. *Int Rev Cytol* 92: 93–132.

Kirkwood TBL. Rosenberger RF, Galas DJ (eds.) (1986) *Accuracy in Molecular Processes: Its Control and Relevance to Living Systems.* Chapman & Hall, London

Kirkwood TBL, Price J, Grove EA (1992) The dispersion of neuronal clones across the cerebral cortex. *Science* 258: 317–320.

Kirkwood TBL, Martin GM, Partridge L (1998) Evolution, senescence, and health in old age. In: *Evolution in Health and Disease* (ed. Stearns SC), pp 219–230, Oxford University Press, Oxford.

Klass MR (1983) A method for the isolation of longevity mutants in the nematode *Caenorhabditis elegans* and initial results. *Mech Ageing Dev* 22: 279–286.

Klein J (1975) *Biology of the Mouse Histocompatibility-2 Complex*, Springer-Verlag, New York.

Klein J (1986) *Natural History of the Major Histocompatibility Complex*, Wiley, New York.

Kobayashi H, Hishida T-O (1992) Electron microscopic observation of an ectopic PGC-like cell in the teleost *Oryzias latipes. Zool Sci* 9: 1087–1092.

Kobyliansky E, Livshits G (1989) Age-dependent changes in morphometric and biochemical traits. *Ann Human Biol* 16: 237–247.

Kohn LAP, Bennett KA (1986) Fluctuating asymmetry in fetuses of diabetic rhesus macaques. *Am J Phys Anthropol* 71: 477–483.

Kopan R, Turner D (1996) The Notch pathway: democracy and aristocracy in the selection of cell fate. *Curr Opin Neurobiol* 6: 594–601.

Kopsidas G, Kovalenko SA, Kelso JM, Linnane AW (1998) An age-associated correlation between cellular bioenergy decline and mtDNA rearrangements in human skeletal muscle. *Mutat Res* 421: 27–36.

Kowald A, Kirkwood TBL (1993a) Explaining fruit fly longevity. *Science* 260: 1664–1665.

Kowald A, Kirkwood TBL (1993b) Mitochondrial mutations, cellular instability and ageing: modelling the population dynamics of mitochondria. *Mutat Res* 295: 93–103.

Kowald A, Kirkwood TBL (1996) A network theory of ageing: the interactions of defective mitochondria, aberrant proteins, free radicals and scavengers in the ageing process. *Mutat Res* 315: 209–236.

Kumar TC (1974) Oogenesis in adult prosimian primates. *Contribut Primatol* 3: 82–96.

Lacalli TC, Harrison LG (1991) From gradients to segments: models for pattern formation in early *Drosophila* embryogenesis. *Seminars Devl Biol.* 2: 107–117.

Lakowski B, Hekimi S (1998) The genetics of caloric restriction in *Caenorhabditis elegans. Proc Natl Acad Sci USA* 95: 13091–13096.

Landfield PW (1994) The role of glucocorticoids in brain aging and Alzheimer's disease: an integrative physiological hypothesis. *Exp Gerontol* 29: 3–11.

Landfield PW, Braun LD, Pitler TA, Lindsey JD, Lynch G (1981) Hippocampal aging in rats: a morphometric study of multiple variables in semithin sections. *Neurobiol Aging* 2: 265–275.

Landfield PW, McEwan BS, Sapolsky RM, Meaney MJ (1996) Hippocampal cell death. *Science* 272: 1249–1251.

Langley SC, Jackson AA (1994) Increased systolic blood pressure in adult rats induced by fetal exposure to maternal low protein diets. *Clin Sci* 86: 217–222.

Lansing AI (1947) A transmissable cumulative and reversible factor in ageing. *J Gerontol* 2: 228–239.

Law CM, Shiell AW (1996) Is blood pressure inversely related to birth weight? The strength of evidence from a systematic review of the literature. *J Hypertens* 14: 935–941.

Lawrence PA (1992) *The Making of a Fly. The Genetics of Animal Design,* Blackwell Scientific Publications, London.

Leamy L (1986) Directional selection and developmental stability: evidence from fluctuating asymmetry of dental characters in mice. *Heredity* 57: 381–388.

Leber SM, Breedlove SM, Sanes JR (1990) Lineage, arrangement, and death of clonally related motoneurons in chick spinal cord. *J Neurosci* 10: 2451–2462.

Lee C (1997) Biology of the prostate ductal system. In: *Prostate: Basic and Clinical Aspects* (ed. Naz RK), pp 53–71, CRC Press, Boca Raton, FL.

Lee MHS, Williams DI (1974) Changes in licking behavior of rat mother following handling of young. *Anim Behav* 22: 679.

LeFevre J, McClintock MK (1988) Reproductive senescence in female rats: a longitudinal study of individual differences in estrous cycles and behavior. *Biol Reprod* 38: 780–789.

Leissner KH, Tisell LE (1979a) The weight of the human prostate. *Scand J Urol Nephrol* 13: 137–142.

Leissner KH, Tisell LE (1979b) The weight of the dorsal, lateral and medial prostatic lobes in man. *Scand J Urol Nephrol* 13: 223–227.

Leistikow EA (1998) Is coronary artery disease initiated perinatally? *Semin Thromb Hemost* 24: 139–143.

LeMaoult J, Delassus S, Dyall R, Nikolic-Zugic J, Kourilsky P, Weksler ME (1997) Clonal expansions of B lymphocytes in old mice. *J Immunol* 159: 3866–3874.

Lengauer C, Kinzler KW, Vogelstein B (1998) Genetic instabilities in human cancers. *Nature* 396: 643–649.

Lerner SP, Finch CE (1991) The major histocompatibility complex and reproductive functions. *Endocr Rev* 12: 78–90.

Lerner SP, Anderson CP, Finch CE (1988) Genotypic influences on female reproductive senescence in mice. *Biol Reprod* 38: 1035–1044.

Lerner SP, Anderson CP, Harrison D, Walford RL, Finch CE (1992) Polygenic influences on the length of oestrus cycles in inbred mice involve MHC alleles. *Eur J Immunogenet* 19: 361–371.

Levay G, Ye Q, Bodell WJ (1997) Formation of DNA adducts and oxidative base damage by copper mediated oxidation of dopamine and 6-hydroxydopamine. *Exp Neurol* 146: 570–574.

LeVay S (1991) A difference in hypothalamic structure between heterosexual and homosexual men. *Science* 253: 1034–1037.

LeVay S (1993) *The Sexual Brain*, MIT Press, Cambridge, Mass.

Leverenz JB, Raskind MA (1998) Early amyloid deposition in the medial temporal lobe of young Down syndrome patients: a regional quantitative analysis. *Exp Neurol* 150: 296–304.

Levin M, Pagan S, Roberts DJ, Cooke J, Kuehn MR, Tabin CJ (1997) Left/right patterning signals and the independent regulation of different aspects of situs in the chick embryo. *Devel Biol* 189: 57–67.

Lewis J (1986) Programs of cell division in the lateral line system. *Trends Neurosci* 9: 135–138.

Li YQ, Roberts SA, Paulus U, Loeffler M, Potten CS (1994) The crypt cycle in mouse small intestinal epithelium. *J Cell Sci* 107: 3271–3279.

Liang P, Hughes V, Fukagawa NK (1997) Increased prevalence of mitochondrial DNA deletions in skeletal muscle of older individuals with impaired glucose tolerance: possible marker of glycemic stress. *Diabetes* 46: 920–923.

Liao A, Nitsch RM, Greenberg SM, Finckh U, Blacker D, Albert M, Rebeck GW, Gomez-Isla T, Clatworthy A, Binetti G, Hock C, Mueller-Thomsen T, Mann U, Zuchowski K, Beisiegel U, Staehelin H, Growdon JH, Tanzi RE, Hyman BT (1998) Genetic association of an alpha2-macroglobulin (Val1000Ile) polymorphism and Alzheimer's disease. *Hum Mol Genet* 7: 1953–1956.

Lin H, Schagat T (1997) Neuroblasts: a model for the asymmetric division of stem cells. *Trends Genet* 13: 33–43.

Lin H, Spradling AC (1993) Germline stem cell division and egg chamber development in transplanted *Drosophila* germaria. *Dev Biol* 159: 140–152.

Lin H, Spradling AC (1997) A novel group of *pumilio* mutations affects the asymmetric division of germline stem cells in the *Drosophila* embryo. *Development* 124: 2463–2476.

Lingner J, Cech TR (1998) Telomerase and chromosome end maintenance. *Curr Opin Genet Dev* 8: 226–232.

Lip GY, Blann AD, Jones AF, Beevers DG (1997) Effects of hormone-replacement therapy on hemostatic factors, lipid factors, and endothelial function in women undergoing surgical menopause: implications for prevention of atherosclerosis. *Am Heart J* 134: 764–771.

Lish JD, Meyer-Bahlburg HF, Ehrhardt AA, Travis BG, Veridiano NP (1992) Prenatal exposure to diethylstilbestrol (DES): childhood play behavior and adult gender-role behavior in women. *Arch Sex Behav* 21: 423–441.

Liu D, Diorio J, Tannenbaum B, Caldji C, Francis D, Freedman A, Sharma S, Pearson D, Plotsky PM, Meaney MJ (1997) Maternal care, hippocampal glucocorticoid receptors, and hypothalamic-pituitary-adrenal responses to stress. *Science* 277: 1659–1662.

Liu VW, Zhang C, Nagley P (1998) Mutations in mitochondrial DNA accumulate differentially in three different human tissues during ageing. *Nucleic Acids Res* 26: 1268–1275.

Livshits G, Kobyliansky E (1989) Age-dependent changes in morphometric and biochemical traits. *Ann Hum Biol* 16: 237–247.

Livshits G, Kobyliansky E (1991) Fluctuating asymmetry as a possible measure of developmental homeostasis in humans: a review. *Human Biol* 63: 441–466.

Ljungquist B, Berg S, Lanke J, McClearn GE, Pedersen NL (1998) The effect of genetic factors for longevity: a comparison of identical and fraternal twins in the Swedish Twin Registry. *J Gerontol A Biol Sci Med Sci* 53: M441–M446.

Loeffler M, Bratke T, Paulus U, Li YQ, Potten CS (1997) Clonality and life cycles of intestinal crypts explained by a state dependent stochastic model of epithelial stem cell organization. *J Theor Biol* 186: 41–54.

Loehlin JC, Martin NG (1998) A comparison of adult female twins from opposite-sex and same-sex pairs on variables related to reproduction. *Behav Genet* 28: 21–27.

Loftus WC, Tramo MJ, Thomas CE, Green RL, Nordgren RA, Gazzaniga MS (1993) Three-dimensional quantitative analysis of hemispheric asymmetry in the human superior temporal region. *Cereb Cortex* 3: 348–355.

Lohr JL, Danos MC, Yost HJ (1997) Left-right asymmetry of a *nodal*-related gene is regulated by dorsoanterior midline structures during *Xenopus* development. *Development* 124: 1465–1472.

Lopera F, Ardilla A, Martinez A, Madrigal L, Arango-Viana JC, Lemere CA, Arango-Lasprilla JC, Hincapie L, Arcos-Burgos M, Ossa JE, Behrens IM, Norton J, Lendon C, Goate AM, Ruiz-Linares A, Rosselli M, Kosik KS (1997) Clinical features of early-onset Alzheimer disease in a large kindred with an E280A presenilin-1 mutation. *J Am Med Assoc* 277: 793–799.

Lowe LA, Supp DM, Sampath K, Yokohama T, Wright CVE, Potter SS, Overbeek P, Kuehn MR (1996) Conserved left-right assymetry of nodal expression and alterations in murine situs invertus. *Nature* 381: 158–161.

Luoh SW, Bain PA, Polakiewicz RD, Goodheart ML, Gardner H, Jaenisch R, Page DC (1997) Zfx mutation results in small animal size and reduced germ cell number in male and female mice. *Development* 124: 2275–2284.

Lynch M, Walsh B (1998) *Genetic Analysis of Quantitative Traits*, Sinauer Associates, Sunderland, Mass.

Macagno ER (1980) Genetic approaches to invertebrate neurogenesis. *Curr Top Dev Biol* 15: 319–345.

Macdonald ME, Ambrose CM, Duyao MP, Myers RH, Lin C, Srinidhi L, Barnes G, Taylor SA, James M, et al. (Huntington Disease Collaborative Research Group) (1993) A novel gene containing a trinucleotide repeat that is expanded and unstable on Huntington-disease chromosomes. *Cell* 72: 971–983.

Madhani HD, Bohr VA, Hanawalt PC (1986) Differential DNA repair in transcriptionally active and inactive proto-oncogenes: *c-abl* and *c-mos*. *Cell* 45: 417–423.

Maini A, Archer C, Wang CY, Haas GP (1997) Comparative pathology of benign prostatic hyperplasia and prostate cancer. *In Vivo* 11: 293–299.

Marcum JB (1974) The freemartin syndrome. *Anim Breeding Abstr* 42: 227–242.

Marín O, González A, Smeets WJAJ (1997) Basal ganglia organization in amphibians: Afferent connections to the striatum and the nucleus accumbens. *J Comp Neurol* 378: 16–49.

Marshall WF, Straight A, Marko JF, Swedlow J, Dernburg A, Belmont A, Murray AW, Agard DA, Sedat JW (1997) Interphase chromosomes undergo constrained diffusional motion in living cells. *Curr Biol* 7: 930–939.

Martin GM, Curtis A, Sprague BS, Epstein CJ (1970) Replicative lifespan of cultivated human cells. Effects of donor's age, tissue and genotype. *Lab Invest* 23: 86–92.

Martin GM, Sprague CA, Norwood TH, Pendergrass WR (1974) Clonal selection, attenuation and differentiation in an *in vitro* model of hyperplasia. *Am J Pathol* 74: 137–154.

Martin GM (1978) Genetic syndromes in man with potential relevance to the pathobiology of aging. In: *Genetic Effects on Aging.* D. Bergsma and D. E. Harrison (eds.), Original Article Ser, Birth Defects, pp. 5–39.

Martin GM (1982) Syndromes of accelerated aging. *Natl Canc Inst Monograph* 60: 241–247.

Martin GM, Austad SN, Johnson TE (1996) Genetic analysis of aging: role of oxidative damage and environmental stresses. *Nat Genet* 13: 25–34.

Martin GM (1997) Genetics and the pathobiology of ageing. *Philos Trans R Soc London [Biol]* 352: 1773–1780.

Martin GM (1998) Atherosclerosis is the leading cause of death in the developed societies. *Am J Pathol* 153: 1319–1320.

Martin K, Kirkwood TBL, Potten CS (1998a) Age changes in stem cells of murine small intestinal crypts. *Exp Cell Res* 241: 316–323.

Martin K, Potten CS, Roberts SA, Kirkwood TBL (1998b) Altered stem cell regeneration in irradiated intestinal crypts of senescent mice. *J Cell Sci* 111: 2297–2303.

Masoro EJ, Austad SN (1996) The evolution of the antiaging action of dietary restriction: a hypothesis. *J Gerontol* 51A: B387–B391.

Mastick CC, Brady MJ, Printen JA, Ribon V, Saltiel AR (1998) Spatial determinants of specificity in insulin action. *Mol Cell Biochem* 182: 65–71.

Mather K (1953) Genetical control of stability in development. *Heredity* 7: 295–336.

Matsumura T, Hayashi M, Konishi R (1985) Immortalization in culture of rat cells: a genealogic study. *J Natl Canc Inst* 74: 1223–1232.

Mayani H, Dragowska W, Lansdorp PM (1993) Lineage commitment in human hemopoiesis involves asymmetric cell division of multipotential progenitors and does not appear to be influenced by cytokines. *J Cell Physiol* 157: 579–586.

Mayeux R, Saunders AM, Shea S, Mirra S, Evans D, Roses AD, Hyman BT, Crain B, Tang MX, Phelps CH (1998) Utility of the apolipoprotein E genotype in the diagnosis of Alzheimer's disease. Alzheimer's Disease Centers Consortium on Apolipoprotein E and Alzheimer's Disease. *N Engl J Med* 338: 506–511.

McAdams HH, Arkin A (1997) Stochastic mechanisms in gene expression. *Proc Natl Acad Sci USA* 94: 814–819.

McAdams HH, Arkin A (1998) Simulation of prokaryotic genetic circuits. *Annu Rev Biophys Biomol Struct* 27: 199–224.

McClaren A (1992) The quest for immortality. *Nature* 359: 482–483.

McClearn GE, Svartengren M, Pedersen NL, Heller DA, Plomin R (1994) Genetic and environmental influences on blood pressure in elderly twins. *Hypertension* 24: 663–670.

McClearn GE, Johansson B, Berg S, Pedersen NL, Ahern F, Petrill SA, Plomin R (1997) Substantial genetic influence on cognitive abilities in twins 80 or more years old. *Science* 276: 1560–1563.

McCrabb GJ, Egan AR, Hosking GJ (1991) Maternal undernutrition during midpregnancy in sheep. Placental size and its relationship to calcium transfer during late pregnancy. *Br J Nutr* 65: 157–168.

McEntee MF, Epstein JI, Syring R, Tierney LA, Strandberg JD (1996) Characterization of prostatic basal cell hyperplasia and neoplasia in aged macaques: comparative pathology in human and nonhuman primates. *Prostate* 29: 51–59.

McEwen BS (1992) Steroid hormones: effect on brain development and function. *Hormone Res* 37 [Suppl 3]: 1–10.

McEwen BS (1997) Hormones as regulators of brain development: life-long effects related to health and disease. *Acta Paediatr [Suppl]* 422: 41–44.

McEwen BS, Sapolsky RM (1995) Stress and cognitive function. *Curr Biol* 5: 205–216.

McFadden DM (1993) A masculinizing effect on the auditory systems of human females having male cotwins. *Proc Natl Acad Sci USA* 90: 11900–11904.

McFadden D, Mishra R (1993) On the relation between hearing sensitivity and otoacoustic emissions. *Hearing Res* 71: 208–213.

McFadden DM, Pasanen EG (1998) Comparison of the auditory systems of heterosexuals and homosexuals: click-evoked otoacoustic emissions. *Proc Natl Acad Sci USA* 95: 2709–2713.

McFadden DM, Pasanen EG, Callaway NL (1998) Changes in otoacoustic emissions in a transsexual male during treatment with estrogen. *J Acoust Soc Am* 104: 1555–1558.

McGue M, Bouchard TJ Jr (1998) Genetic and environmental influences on human behavioral differences. *Annu Rev Neurosci* 21: 1–24.

McKeithan TW (1995) Kinetic proofreading in T-cell receptor signal transduction. *Proc Natl Acad Sci USA* 92: 5042–5046.

McKenzie JA, Batterham P (1994) The genetic, molecular, and phenotypic consequences of selection for insecticide resistance. *Trends Ecol Evol* 9: 166–169.

McKenzie JA, Yen JL (1995) Genotype, environment and the asymmetry phenotype. Dieldrin-resistance in *Lucilia cuprina* (the Australian sheep blowfly). *Heredity* 75: 181–187.

McLachlan JA, Newbold RR, Bullock BC (1975) Reproductive tract lesions in male mice exposed prenatally to diethylstilbestrol. *Science* 190: 991–992.

McNeal JE (1978) Origin and evolution of benign prostatic enlargement. *Invest Urol* 15: 340–345.

McNeal JE (1990) Pathology of benign prostatic hyperplasia. Insight into etiology. *Urol Clin N Am* 17: 477–486.

Meaney MJ, Aitken DH, van Berkel C, Bhatnagar S, Sapolsky RM (1988) Effect of neonatal handling on age-related impairments associated with the hippocampus. *Science* 239: 766–768.

Medawar PB (1952) *An Unsolved Problem of Biology*, Lewis, London.

Meikle AW, Stephenson RA, McWhorter WP, Skolnick MH, Middleton RG (1995) Effects of age, sex steroids, and family relationships on volumes of prostate zones in men with and without prostate cancer. *Prostate* 26: 253–259.

Meikle AW, Stephenson RA, Lewis CM, Middleton RG (1997a) Effects of age and sex hormones on transition and peripheral zone volumes of prostate and benign prostatic hyperplasia in twins. *J Clin Endoc Metab* 82: 571–575.

Meikle AW, Stephenson RA, Lewis CM, Wiebke GA, Middleton RG (1997b) Age, genetic, and nongenetic factors influencing variation in serum sex steroids and zonal volumes of the prostate and benign prostatic hyperplasia in twins. *Prostate* 33: 105–111.

Meikle AW, Bansal A, Murray DK, Stephenson RA, Middleton RG (1999) Heritability of the symptoms of benign prostatic hyperplasia and the roles of age and zonal prostate volumes in twins. *Urology* 53: 701–706.

Meikle DB, Thornton MW (1995) Premating and gestational effects of maternal nutrition on secondary sex ratio in house mice. *J Reprod Fertil* 105: 193–196.

Melov S, Hertz GZ, Stormo GD, Johnson TE (1994) Detection of deletions in the mitochondrial genome of *Caenorhabditis elegans*. *Nucleic Acids Res* 22: 1075–1078.

Melov S, Shoffner JM, Kaufman A, Wallace DC (1995) Marked increase in the number and variety of mitochondrial DNA rearrangements in aging human skeletal muscle. *Nucleic Acids Res* 23: 4122–4126.

Meyer JM, Breitner JC (1998) Multiple threshold model for the onset of Alzheimer's disease in the NAS-NRC twin panel. *Am J Med Genet* 81: 92–97.

Meyer MR, Tschanz JT, Norton MC, Welsh-Bohmer KA, Steffens DC, Wyse BW, Breitner JC (1998) APOE genotype predicts when—not whether—one is predisposed to develop Alzheimer disease. *Nat Genet* 19: 321–322.

Meyers-Wallen VN (1993) Genetics of sexual differentiation and anomalies in dogs and cats. *J Reprod Fertil [Suppl]* 47: 441–452.

Micklem HS, Ross E (1978) Heterogeneity and ageing of haematopoietic stem cells. *Ann Immunol (Inst Pasteur)* 129C: 367–376.

Miki Y (1998) Retrotransposal integration of mobile genetic elements in human diseases. *J Hum Genet* 43: 77–84.

Mizutani T, Shimada H (1992) Neuropathological background of twenty-seven centenarian brains. *J Neurol Sci* 108: 168–177.

Mobbs CV, Flurkey K, Gee DM, Yamamoto K, Sinha YN, Finch CE (1984) Estradiol-induced adult anovulatory syndrome in female C57BL/6J mice: Age-like neuroendocrine, but not ovarian, impairments. *Biol Reprod* 30: 556–563.

Mohr C, Görner P (1996) Innervation patterns of the lateral line stiches of the clawed frog, *Xenopus laevis*, and their reorganization during metamorphosis. *Brain Behav Evol* 48: 55–69.

Møller AP (1996a) Sexual selection, viability selection, and developmental stability in the domestic fly *Musca domestica*. *Evolution* 50: 746–752.

Møller AP (1996b) Development of fluctuating asymmetry in tail feathers of the barn swallow *Hirundo rustica*. *J Evol Biol* 9: 677–694.

Møller AP, Swaddle JP (1997) *Asymmetry, Developmental Stability, and Evolution*, Oxford Series in Ecology and Evolution, Oxford Univ Press, New York.

Møller AP, Soler M, Thornhill R (1995) Breast asymmetry, sexual selection, and human reproductive success. *Ethol Sociobiol* 16: 207–219.

Moran S, Ren RXF, Kool ET (1997) A thymidine triphosphate shape analog lacking Watson-Crick pairing ability is replicated with high sequence selectivity. *Proc Natl Acad Sci USA* 94: 10506–10511.

Morgan DG, Marcusson JO, Nyberg P, Wester P, Winblad BN, Gordon MN, Finch CE (1987) Divergent changes in D-1 and D-2 dopamine binding sites of human brain during aging. *Neurobiol Aging* 8: 195–201.

Morgan TE, Xie Z, Goldsmith S, Yoshida T, Lanzrein A-S, Stone D, Rozovsky I, Perry G, Smith MA, Finch CE (1999) The mosaic of brain glial hyperactivity during normal aging and its attenuation by food restriction. *Neuroscience* 89: 687–699.

Morimoto Y (1998) Estimation of the branching number of the blood vessel system. *Physiol Lett A* 242: 285–289.

Morley AA, Cox S, Holliday R (1982) Human lymphocytes resistant to 6-thioguanine increase with age. *Mech Ageing Dev* 19: 21–26.

Morris JC, Storandt M, McKeel DW Jr, Rubin EH, Price JL, Grant EA, Berg L (1996) Cerebral amyloid deposition and diffuse plaques in "normal" aging: evidence for presymptomatic and very mild Alzheimer's disease. *Neurology* 46: 707–719.

Morrison SJ, Wandycz AM, Akashi K, Globerson A, Weissman (1996) The aging of hematopoietic stem cells. *Nat Med* 2: 1011–1016.

Morwessel NJ (1998) The genetic basis of diabetes mellitus. *AACN Clin Issues* 9: 539–554.

Murphy JS, Davidoff M (1972) The result of improved nutrition on the Lansing effect in *Moina macrocopa. Biol Bull* 142: 302–309.

Nagamani M, McDonough PG, Ellegood JO, Mahesh VB (1979) Maternal and amniotic fluid steroids throughout human pregnancy. *Am J Obstet Gynecol* 134: 674–680.

Nagel SC, vom Saal FS, Thayer KA, Dhar MG, Boechler M, Welsohons WV (1997) Relative binding affinity-serum modified access (RBA-SMA) assay predicts the relative in vivo bioactivity of the xenoestrogens bisphenol A and octylphenol. *Environ Health Perspect* 105: 70–76.

Namba R, Pazdera TM, Cerrone RL, Minden JS (1997) *Drosophila* embryonic pattern repair: how embryos respond to *bicoid* dosage alteration. *Development* 124: 1393–1403.

Napoli C, D'Armiento FP, Mancini FP, Postiglione A, Witztum JL, Palumbo G, Palinski W (1997) Fatty streak formation occurs in human fetal aortas and is greatly enhanced by maternal hypercholesterolemia. Intimal accumulation of low density lipoprotein and its oxidation precede monocyte recruitment into early atherosclerotic lesions. *J Clin Invest* 100: 2680–2690.

Nasr A, Breckwoldt M (1998) Review: estrogen replacement therapy and cardiovascular protection: lipid mechanisms are the tip of an iceberg. *Gynecol Endocrinol* 12: 43–59.

Nayar SDO, Reggio BBS, Henson M (1996) Association of endogenous steroid levels with high and low density lipoprotein concentrations in the human maternal-fetoplacental unit at term. *J Reprod Med* 41: 91–98.

Neale MC, Cardon LR (1992) *Methodology for Genetic Studies of Twins and Families*, NATO ASI Series, Series D: Behavioral and Social Sciences, vol 67, Kluwer, Boston.

Neel JV (1962) DM: a "thrifty" genotype rendered detrimental by progress? *Am J Hum Genet* 14: 352–362.

Nelson JF, Felicio LS (1986) Radical ovarian resection advances the onset of persistent vaginal cornification, but only transiently disrupts hypothalamic-pituitary regulation of cyclicity in C57BL/6J mice. *Biol Reprod* 35: 957–964.

Nelson JF, Felicio LS, Osterburg HH, Finch CE (1981) Altered profiles of estradiol and progesterone associated with prolonged estrous cycles and persistent vaginal cornification in aging C57BL/6J mice. *Biol Repro* 24: 784–794.

Nelson JF, Felicio LS, Randall PK, Simms C, Finch CE (1982) A longitudinal study of estrous cyclicity in aging C57BL/6J mice: 1, cycle frequency, length and vaginal cytology. *Biol Reprod* 27: 327–339.

Nelson JF, Gosden RG, Felicio LS (1985) Effect of dietary restriction on estrous cyclicity and follicular reserves in aging C57BL/6J mice. *Biol Reprod* 32: 515–522.

Neveu PJ (1993) Brain lateralization and immunomodulation. *Int J Neurosci* 70: 135–143.

Newbold RR, McLachlan JA (1982) Vaginal adenosis and adenocarcinoma in mice exposed prenatally or neonatally to diethylstilbestrol. *Cancer Res* 42: 2003–2011.

Newbold RR, McLachlan JA (1996) Transplacental hormonal carcinogenesis: diethylstilbestrol as an example. *Prog Clin Biol Res* 394: 131–147.

Newbold RR, Bullock BC, McLachlan JA (1983a) Exposure to diethylstilbestrol during pregnancy permanently alters the ovary and oviduct. *Biol Reprod* 28: 735–744.

Newbold RR, Tyrey S, Haney AF, McLachlan JA (1983b) Developmentally arrested oviduct: a structural and functional defect in mice following prenatal exposure to diethylstilbestrol. *Teratology* 27: 417–426.

Newbold RR, Bullock BC, McLachlan JA (1985) Progressive proliferative changes in the oviduct of mice following developmental exposure to diethylstilbestrol. *Teratogen Carcinogen Mutagen* 5: 473–480.

Newbold RR, Bullock BC, McLachlan JA (1987) Müllerian remnants of male mice exposed prenatally to diethylstilbestrol. *Teratogen Carcinogen Mutagen* 7: 377–389.

Nguyen TV, Howard GM, Kelly PJ, Eisman JA (1998) Bone mass, lean mass, and fat mass: same genes or same environments? *Am J Epidemiol* 147: 3–16.

Nickerson JA, Blencow BJ, Penman S (1995) The architectural organization of nuclear metabolism. *Int Rev Cytol* 162A: 67–123.

Nicklas RB (1997) How cells get the right chromosomes. *Science* 275: 632–637.

Ninio J (1975) Kinetic amplification of enzyme discrimination. *Biochimie* 57: 587–595.

Ninio J (1986) Kinetic and probabilistic thinking in accuracy. In: *Accuracy in Molecular Processes* (ed. Kirkwood TBL, Rosenberger RF, Galas DJ), pp 291–328, Chapman and Hall, London.

Ninio J (1987) Kinetic devices in protein synthesis, DNA replication, and mismatch repair. *Cold Spring Harbor Symp Quant Biol* 52: 639–646.

Nittenberg R, Patel K. Joshi Y, Krumlauf R, Wilkinson DG, Brickell PM, Tickle C, Clarke JD (1997) Cell movements, neuronal organisation and gene expression in hindbrains lacking morphological boundaries. *Development* 124: 2297–2306.

Noble EP, Berman SM, Ozkaragoz TZ, Ritchie T (1994) Prolonged P300 latency in children with the D2 dopamine receptor A1 allele. *Am J Hum Genet* 54: 658–668.

Norman RL, Spies HG (1986) Cyclic ovarian function in a male macaque: additional evidence for a lack of sexual differentiation in the physiological mechanisms that regulate the cyclic release of gonadotropins in primates. *Endocrinology* 118: 2608–2610.

Nottebohm F (1996) The King Solomon Lectures in Neuroethology. A white canary on Mount Acropolis. *J Comp Physiol [A]* 179: 149–156.

Oakley G-P (1997) Doubling the number of women consuming vitamin supplement pills containing folic acid: an urgently needed birth defect prevention complement to the folic acid fortification of cereal grains. *Toxicology* 11: 579–581.

Oakley G-P Jr, Adams MJ, Dickinson CM (1996) More folic acid for everyone, now. *J Nutrit* 126: 751S–755S.

Odagiri Y, Uchida H, Hosokawa M, Takemoto K, Morley AA, Takeda T (1998) Accelerated accumulation of somatic mutations in the senescence-accelerated mouse. *Nat Genet* 19: 116–117.

Ogden DA, Micklem HS (1976) The fate of serially transplanted bone marrow cell populations from young and old donors. *Transplantation* 22: 287–293.

Ohkubo Y, Shirayoshi Y, Nakatsuji N (1996) Autonomous regulation of proliferation and growth arrest in mouse primordial germ cells studied by mixed and clonal cultures. *Exp Cell Res* 222: 291–297.

Ohta H, Sugimoto I, Masuda A, Komukai S, Suda Y, Makita K, Takamatsu K, Horiguchi F, Nozawa S (1996) Decreased bone mineral density associated with early menopause progresses for at least ten years: cross-sectional comparisons between early and normal menopausal women. *Bone* 18: 227–231.

Okuizumi K, Onodera O, Namba Y, Ikeda K, Yamamoto T, Seki K, Ueki A, Nanko S, Tanaka H, Takahashi H, et al. (1995) Genetic association of the very low density lipoprotein (VLDL) receptor gene with sporadic Alzheimer's disease. *Nature Genet* 11: 207–209.

Oppenheim JS, Skerry JE, Tramo MJ, Gazzinaga MS (1989) Magnetic resonance imaging morphology of the corpus callosum in monozygotic twins. *Ann Neurol* 26: 100–104.

Oppenheim RW (1989) Early regional variations in motoneuron numbers arise by differential proliferation in the chick embryo spinal cord. *Dev Biol* 133: 468–474.

Oppenheim RW (1991) Cell death during development of the nervous system. *Annu Rev Neurosci* 14: 453–501.

Paganelli GM, Santucci R, Biasco A, Miglioli M, Barbara L (1990) Effect of age and sex on rectal cell renewal in humans. *Cancer Lett* 53: 117–121.

Paganini-Hill A, Henderson VW (1994) Estrogen deficiency and risk of Alzheimer's disease in women. *Am J Epidemiol* 140: 256–261.

Page WF, Braun MM, Partin AW, Caporaso N, Walsh P (1997) Heredity and prostate cancer: a study of World War II veteran twins. *Prostate* 33: 240–245.

Pakkenberg B, Gundersen HJG (1997) Neocortical neuron number in humans: effects of sex and age. *J Comp Neurol* 384: 312–320.

Palanza P, Parmigiani S, vom Saal FS (1995) Urine marking and maternal aggression of wild female mice in relation to anogenital distance at birth. *Physiol Behav* 58: 827–835.

Palmeirim I, Henrique D, Ish-Horowicz D, Pourquie O (1997) Avian hairy gene expression identifies a molecular clock linked to vertebrate segmentation and somitogenesis. *Cell* 91: 639–648.

Palmer RA, Strobeck C (1986) Fluctuating asymmetry: measurement, analysis, patterns. *Annu Rev Ecol Syst* 17: 391–421.

Parkes TL, Elia AJ, Dickinson D, Hilliker AJ, Phillips JP, Boulianne GL (1998) Extension of *Drosophila* lifespan by overexpression of human SOD1 in motoneurons. *Nat Genet* 19: 171–174.

Parsons PA (1990) Fluctuating asymmetry: an epigenetic measure of stress. *Biol Rev* 65: 131–145.

Parsons PA (1992) Fluctuating asymmetry: a biological monitor of environmental and genomic stress. *Heredity* 68: 361–364.

Partin AW, Page WF, Lee BR, Sanda MG, Miller RN, Walsh PC (1994) Concordance rates for benign prostatic disease among twins suggest hereditary influence. *Urology* 44: 646–650.

Pedersen JF, Mølsted-Pedersen L (1981) Early fetal growth delay detected by ultrasound marks increased risk of congenital malformation in diabetic pregnancy. *Br Med J* 283: 269–270.

PEPI Trial Writing Group (1995) Effects of estrogen/progestin regimens on heart disease risk factors in postmenopausal women. The Postmenopausal Estrogen/Progestin Interventions (PEPI) Trial. *J Am Med Assoc* 273: 199–208.

Pepys MB, Dash AC, Markham RE, Thomas HC, Williams BD, Petrie A (1978) Comparative clinical study of protein SAP (amyloid P component) and C-reactive protein in serum. *Clin Exp Immunol* 32: 119–124.

Perez GI, Robles R, Knudson CM, Flaws JA, Korsmeyer SJ, Tilly JL (1999) Prolongation of ovarian lifespan into advanced chronological age by Bax-deficiency. *Nature Genet* 21: 200–203.

Perry VH, Matyszak MK, Fearn S (1993) Altered antigen expression of microglia in the aged rodent CNS. *Glia* 7: 60–67.

Pesce M, Farrqace MG, Piacentini M, Dolci S, De Felici M (1993) Stem cell factor and leukemia inhibitory factor promote primordial germ cell survival by suppressing programmed cell death (apoptosis). *Development* 118: 1089–1094.

Petruska J, Goodman MF (1995) Enthalpy-entropy compensation in DNA melting thermodynamics. *J Biol Chem* 270: 746–750.

Phelan JP, Austad SN (1994) Selecting animal models of human aging: inbred strains often exhibit less biological uniformity than F_1 hybrids. *J Gerontol* 49: B1–B11.

Phillips DIW (1993) Twin studies in medical research: can they tell us whether diseases are genetically determined? *Lancet* 341: 1008–1009.

Pienta KJ, Esper PS (1993) Risk factors for prostate cancer. *Ann Int Med* 118: 793–803.

Pike MC, Ursin G (1994) Etiology of benign prostatic hyperplasia: can this disease be prevented? In: *Benign Prostatic Hyperplasia. Innovations in Management* (ed. Petrovich Z, Baert L), pp 1–16, Springer-Verlag, New York.

Pillard RC, Bailey JM (1998) Human sexual orientation has a heritable component. *Hum Biol* 70: 347–365.

Pillard RC, Rosen LR, Meyer-Bahlburg H, Weinrich JD, Feldman JF, Gruen R, Ehrhardt AA (1993) Psychopathology and social functioning in men prenatally exposed to diethylstilbestrol (DES). *Psychosom Med* 55: 485–491.

Pincus S, Singer BH (1998) A recipe for randomness. *Proc Natl Acad Sci USA* 95: 10367–10372.

Plassman BL, Welsh-Bohmer KA, Bigler ED, Johnson SC, Anderson CV, Helms MJ, Saunders AM, Breitner JC (1997) Apolipoprotein E ε4 allele and hippocampal volume in twins with normal cognition. *Neurology* 48: 985–989.

Pletcher SD, Curtsinger JW (1998) Mortality plateaus and the evolution of senescence: why are the old-age mortality rates so low? *Evolution* 52: 454–464.

Plomin R (1994) *Genetics and Experience. The Interplay between Nature and Nurture*, Sage, Thousand Oaks, Cali.

Plomin R, Daniels D (1987) Why are children in the same family so different from one another? *Behav Brain Sci* 10: 1–60.

Poirie M, Niederer E, Steinmann-Zwicky M (1995) A sex-specific number of germ cells in embryonic gonads of *Drosophila*. *Development* 121: 1867–1873.

Polak M (1997) Parasites, fluctuating asymmetry, and sexual selection. In: *Parasites and Pathogens. Effects of Host Hormones and Behavior*, chap 13, pp 246–276, Chapman and Hall, New York.

Pontén J, Stolt L (1980) Proliferation control in cloned normal and malignant human cells. *Exp Cell Res* 129: 367–375.

Pontén J, Stein WD, Shall S (1983) Quantitative analysis of the aging of human glial cells in culture. *J Cell Physiol* 117: 342–352.

Poole TB, Evans RG (1982) Reproduction, infant survival and productivity of a colony of common marmosets (*Callithrix jacchus jacchus*). *Lab Anim* 16: 88–97.

Potten CS (1992) The significance of spontaneous and induced apoptosis in the gastrointestinal tract of mice. *Cancer Metastasis Rev* 11: 179–195.

Potten CS (1998) Stem cells in gastrointestinal epithelium: numbers, characteristics and death. *Philos Trans R Soc Lond [Biol]* 353: 821–830.

Potten CS, Hendry JH (1995) Clonal regeneration studies. In: *Radiation and Gut* (ed. Potten CS, Hendry JH), pp 45–59, Elsevier, Amsterdam.

Potten CS, Loeffler M (1990) Stem cells: attributes, cycles, spirals, pitfalls and uncertainties. Lessons for and from the crypt. *Development* 110: 1001–1020.

Potten CS, Booth C, Pritchard DM (1997) The intestinal epithelial stem cell: the mucosal governor. Stem cell review. *Int J Expl Pathol* 78: 219–243.

Pratley RE (1998) Gene-environment interactions in the pathogenesis of type 2 diabetes mellitus: lessons learned from the Pima Indians. *Proc Nutr Soc* 57: 175–181.

Price DL, Sisidia SS, Borchelt DR (1998) Alzheimer disease—when and why. *Nat Genet* 19: 314–316.

Price DL, Martin LJ, Sisidia SS, Walker LC, Cork LC (1992) Alzheimer's disease-type brain abnormalities in animal models. *Prog Clin Biol Res* 379: 271–287.

Price T, Schluter D (1991) On the low heritability of life-history traits. *Evolution* 45: 853–861.

Pries AR, Secomb TW, Gaehtgens P (1996) Relationship between structural and hemodynamic heterogeneity in microvascular networks. *Am Physiol Soc* 270: H545–H553.

Prins GS, Jung MH, Vellanoweth RL, Chatterjee B, Roy AK (1996) Age-dependent expression of the androgen receptor gene in the prostate and its implication in glandular differentiation and hyperplasia. *Dev Genet* 18: 99–106.

Probst R, Lonsbury-Martin BL, Martin GK (1991) A review of otoacoustic emissions. *J Acoust Soc Am* 89: 2027–2067.

Prodohl PA, Loughry WJ, McDonough CM, Nelson WS, Avise JC (1996) Molecular documentation of polyembryony and the micro-spatial dispersion of clonal sibships in the nine-banded armadillo *Dasypus novemcinctus, Proc R Soc Lond [Biol]* 263: 1643–1649.

Promislow DE, Tatar M, Khazaeli AA, Curtsinger JW (1996) Age-specific patterns of genetic variance in *Drosophila melanogaster*. I. Mortality. *Genetics* 143: 839–848.

Rabinovici J, Jaffe RB (1990) Development and regulation of growth and differentiated function in human and subhuman primate fetal gonads. *Endocr Rev* 11: 532–557. [Erratum appears in *Endocr Rev* (1991) 12: 90.]

Rafferty KA Jr (1985) Growth potential of the cells of permanent lines (HeLa, BHK/21, NRK). *Virchows Arch [Cell Pathol]* 50: 167–180.

Rahmouni A, Yang A, Tempany CM, Frenkel T, Epstein J, Walsh P, Leichner PK, Ricci C, Zerhouni E (1992) Accuracy of in vivo assessment of prostatic volume by MRI and transrectal ultrasonography. *J Comput Assist Tomogr* 16: 935–940.

Rakic P (1972) Mode of cell migration to the superficial layers of fetal monkey neocortex. *J Comp Neurol* 145: 61–83.

Rakic P (1991) Experimental manipulation of cerebral cortical areas in primates. *Philos Trans R Soc Lond [Biol]* 331: 291–294.

Randerath K, Putnam KL, Osterburg HH, Johnson SA, Morgan DG, Finch CE (1993) Age-dependent increases of DNA adducts (I-compounds) in human and rat brain DNA. *Mutat Res* 295: 11–18.

Ranganathan R, Ross EM (1997) PDZ domain proteins: scaffolds for signaling complexes. *Curr Biol* 7: R770–R773.

Rasmussen T, Schliemann T, Sorensen JC, Zimmer J, West MJ (1996) Memory impaired aged rats: no loss of principal hippocampal and subicular neurons. *Neurobiol Aging* 17: 143–147.

Ratcliffe MA, Lanham SA, Reid DM, Dawson AA (1992) Bone mineral density (BMD) in patients with lymphoma: the effects of chemotherapy, intermittent corticosteroids and premature menopause. *Hematol Oncol* 10: 181–187.

Ravelli ACJ, van der Meulen JHP (1998) Prenatal exposure to famine and health in later life. Authors' reply. *Lancet* 351: 1362.

Ravelli ACJ, van der Meulen JHP, Michels RPJ, et al. (1998) Glucose tolerance in adults after prenatal exposure to famine. *Lancet* 35: 173–177.

Ray WJ, Yao M, Nowotny P, Mumm J, Zhang W, Wu JY, Kopan R, Goate AM (1999) Evidence for a physical interaction between presenilin and Notch. *Proc Natl Acad Sci USA* 96: 3263–3268.

Rea IM, Middleton D (1994) Is the phenotypic combination A1B8Cw/7DR3 a marker for male longevity? *J Am Geriatr Soc* 42: 978–983.

Real LA (1980) On uncertainty and the law of diminishing returns in evolution and behavior. In: *Limits to Action* (ed. Staddon JER), pp 37–64, Academic Press, New York.

Reeve J, Kroger H, Nijs J, Pearson J, Felsenberg D, Reiners C, Schneider P, Mitchell A, Ruegsegger P, Zander C, Fischer M, Bright J, Henley M, Lunt M, Dequeker J (1996) Radial cortical and trabecular bone densities of men and women standardized with the European forearm phantom. *Calcif Tissue Int* 58: 135–143.

Reiman EM, Caselli RJ, Yun LS, Chen K, Bandy D, Minoshima S, Thibodeau SN, Osborne D (1996) Preclinical evidence of Alzheimer's disease in persons homozygous for the epsilon 4 allele for apolipoprotein E. *N Engl J Med* 334: 752–758.

Resch A, Schneider B, Bernecker P, Battmann A, Wergedal J, Willvonseder R, Resch H (1995) Risk of vertebral fractures in men: relationship to mineral density of the vertebral body. *Am J Roentgenol* 164: 1447–1450.

Resko JA, Roselli CE (1997) Prenatal hormones organize sex differences of the neuroendocrine reproductive system: observations on guinea pigs and non-human primates. *Cell Mol Neurobiol* 17: 627–648.

Resko JA, Ellinwood WE, Pasztor LM, Buhl AE (1980) Sex steroids in the umbilical circulation of fetal rhesus monkeys from the time of gonadal differentiation. *J Clin Endocr Metab* 50: 900–905.

Reyes FI, Boroditsky RS, Winter JSD, Faiman C (1974) Studies on human sexual development II, fetal and material serum gonadotrophin and sex steroid concentrations. *J Clin Endocrinol Metab* 38: 612–617.

Reznikov KY (1991) *Cell Proliferation and Cytogenesis in the Mouse Hippocampus*, Springer-Verlag, Berlin.

Richardson SJ, Senikas V, Nelson JF (1987) Follicular depletion during the menopausal transition: evidence for accelerated loss and ultimate exhaustion. *J Clin Endocrinol Metab* 65: 1231–1237.

Richardson SJ, Senikas V, Nelson JF (1988) Follicular depletion during the menopausal transition: evidence for accelerated loss and ultimate exhaustion. *J Clin Endocrinol Metab* 65: 1231–1240.

Rich-Edwards JW, Stampfer MJ, Manson JE, Rosner B, Hankinson SE, Colditz GA, Willett WC, Hennekens CH (1997) Birth weight and risk of cardiovascular disease in a cohort of women followed up since 1976. *Brit Med J* 315: 396–400.

Riggs BL, Melton LJ 3d (1986) Involutional osteoporosis. *N Engl J Med* 314: 1676–1686.

Rines JP, vom Saal FS (1984) Fetal effects on sexual behavior and aggression in young and old female mice treated with estrogen and testosterone. *Horm Behav* 18: 117–129.

Ritchie K (1995) Mental status examination of an exceptional case of longevity J. C. aged 118 years. *Br J Psychiatry* 166: 229–235.

Rivera-Pomar R, Jäckle H (1996) From gradients to stripes in *Drosophila* embryogenesis: filling in the gaps. *Trends Genet* 12: 478–483.

Rivière I, Sunshine MJ, Littman DR (1998) Regulation of IL-4 expression by activation of individual alleles. *Immunity* 9: 217–228.

Roberts SA, Hendry JH, Potten CS (1995) Deduction of the clonogen content of intestinal crypts: a direct comparison of two-dose and multiple-dose methodologies. *Radiat Res* 141: 303–308.

Rodeck CH, Gill D, Rosenberg DA, Collins WP (1985) Testosterone levels in midtrimester maternal and fetal plasma and amniotic fluid. *Prenat Diag* 5: 175–181.

Roff DA, Mousseau TA (1987) Quantitative genetics and fitness: Lessons from *Drosophila*. *Heredity* 58: 103–118.

Röhme D (1981) Evidence for a relationship between longevity of mammalian species and lifespans of normal fibroblasts *in vitro* and erythrocytes *in vivo*. *Proc Natl Acad Sci USA* 78: 5009–5013.

Roncucci L, Ponz de Leon M, Scalmati A, Malagoli G, Pratissoli S, Perini M, Chahin NJ (1988) The influence of age on colonic epithelial cell proliferation. *Cancer* 62: 2373–2377.

Rose MR (1991) *The Evolutionary Biology of Aging*, Oxford University Press, Oxford.

Rosenberger RF (1991) Senescence and the accumulation of abnormal proteins. *Mutat Res* 256: 255–262.

Roses AD (1997) Apolipoprotein E, a gene with complex biological interactions in the aging brain. *Neurobiol Dis* 4: 170–185.

Roses AD, Saunders AM (1997) Apolipoprotein E genotyping as a diagnostic adjunct for Alzheimer's disease. *Int Psychogeriatr* 9 [Suppl 1]: 277–288.

Ross IL, Browne CM, Hume DA (1994) Transcription of individual genes in eukaryotic cells occurs randomly and infrequently. *Immunol Cell Biol* 72: 177–185.

Rothenberg EV, Telfer JC, Anderson MK (1999) Transcriptional regulation of lymphocyte lineage commitment. *BioEssays*, in press.

Rothstein M (1982) *Biochemical Approaches to Aging*, Academic Press, New York.

Rubelc I, Vondraĉek Z (1999) Stochastic mechanisms of cellular aging: abrupt telomere shortening as a model for the stochastic nature of cellular aging. *J Theor Biol* 197: 425–438.

Rubin H (1997) Cell aging in vivo and in vitro. *Mech Ageing Dev* 98: 1–35.

Rubinsztein DC, Leggo J, Coles R, Almqvist E, Biancalana V, Cassiman JJ, Chotai K, Connarty M, Crauford D, Curtis A, Curtis D, Davidson MJ, Differ AM, Dode C, Dodge A, Frontali M, Ranen NG, Stine OC, Sherr M, Abbott MH, Franz ML, Graham CA, Harper PS, Hedreen JC, Hayden MR, et al. (1996) Phenotypic characterization of individuals with 30–40 CAG repeats in the Huntington disease (HD) gene reveals HD cases with 36 repeats and apparently normal elderly individuals with 36–39 repeats. *Am J Hum Genet* 59: 16–22.

Rudolph KL, Chang S, Lee HW, Blasco M, Gottlieb GJ, Greider C, DePinho RA (1999) Longevity, stress response, and cancer in aging telomerase-deficient mice. *Cell* 96: 701–712.

Russo I, Silver AR, Cuthebert AP, Griffin DK, Trott DA, Newbold RF (1998) A telomere-independent senescence mechanism is the sole barrier to Syrian hamster cell immortalization. *Oncogene* 17: 3417–3426.

Ryan AK, Blumberg B, Rodriguez-Esteban C, Yonei-Tamura S, Tamura K, Tsukui T, Pena J, Sabbagh W, Greenwald J, Choe S, Norris DP, Robertson EJ, Evans RM, Rosenfeld MG, Belmonte JCI (1998) Pitx2 determines left–right asymmetry of internal organs in vertebrates. *Nature* 394: 545–551.

Ryan JM, Ostrow DG, Breakefield XO, Gershon ES, Upchurch L (1981) A comparison of the proliferative and replicative life span kinetics of cell cultures derived from monozygotic twins. *In Vitro* 17: 20–27.

Sako Y, Kasumi A (1994) Compartmentalized structure of the plasma membrane for receptor movements as revealed by a nanometer-level motion analysis. *J Cell Biol* 125: 1251–1264.

Sanda MG, Beatty TH, Stutzman RE, Childs B, Walsh PC (1994) Genetic susceptibility of BPH. *J Urol* 152: 115–119.

Sapolsky RM (1996) Why stress is bad for your brain. *Science* 273: 749–750.

Sapolsky RM (1997) The importance of a well-groomed child. *Science* 277: 1620–1621.

Sapolsky RM, Uno H, Rebert CS, Finch CE (1990) Hippocampal damage associated with prolonged glucocorticoid exposure in primates. *J Neurosci* 10: 2897–2902.

Saunders JW (1966) Death in embryonic systems. *Science* 254: 604–612.

Schächter F, Faure-Delanef L, Guenot F, Rouger H, Froguel P, Lesueur-Gginot L, Cohen D (1994) Genetic associations with human longevity at the APOE and ACE loci. *Nat Genet* 6: 29–32.

Scharf J, Dovrat A, Gershon D (1987) Defective superoxide-dismutase molecules accumulate with age in human lenses. *Graefes Arch Clin Exp Ophthalmol* 225: 133–136.

Schipper H, Brawer JR, Nelson JF, Felicio LS, Finch CE (1981) Role of the gonads in the histologic aging of hypothalamic arcuate nucleus. *Biol Reprod* 25: 413–419.

Schipper HE, Stopa EG (1995) Expression of heme oxygenase-1 in the senescent and Alzheimer-diseased brain. *Ann Neurol* 37: 758–768.

Schneider EL, Y Mitsui Y (1976) The relationship between in vitro cellular aging and in vivo human age. *Proc Natl Acad Sci USA* 73: 3584–3588.

Schneider LS, Finch CE (1997) Can estrogen prevent neurodegeneration? *Drugs Aging* 11: 87–95.

Schneider LS, Farlow MR, Henderson VW, Pogoda JM (1996) Effects of estrogen replacement therapy on response to tacrine in patients with Alzheimer's disease. *Neurology* 46: 1580–1584.

Schupf N, Kapell D, Nightingale B, Rodriguez A, Tycko B, Mayeux R (1998) Earlier onset of Alzheimer's disease in men with Down syndrome. *Neurology* 50: 991–995.

Schwabl H (1993) Yolk is a source of maternal testosterone for developing birds. *Proc Natl Acad Sci USA* 90: 11446–11450.

Schwabl H (1996) Environment modifies the testosterone level of a female bird and its eggs. *J Exp Zool* 276: 157–163.

Schweisguth F, Gho M, Lecourtois M (1996) Control of cell fate choices by lateral signaling in the adult peripheral nervous system of *Drosophila melanogaster*. *Dev Genet* 18: 28–39.

Scrimshaw NS (1997) The relation between fetal malnutrition and chronic disease in later life. *Br Med J* 315: 825–826.

Searle J, Kerr JFR, Bishop CJ (1982) Necrosis and apoptosis: distinct modes of cell death with fundamentally different significance. *Pathol Annu* 17: 229–259.

Seeman TE, McEwen BS, Singer BH, Albert MS, Rowe JW (1997) Increase in urinary cortisol excretion and memory declines: MacArthur studies of successful aging. *J Clin Endocrinol Metab* 82: 2458–2465.

Seldin MF, Conroy J, Steinberg AD, D'Hoosteleare LA, Raveche ES (1987) Clonal expansion of abnormal B cells in old NZB mice. *J Exp Med* 166: 1585–1590.

Selkoe DJ (1994) Normal and abnormal biology of the beta-amyloid precursor protein. *Annu Rev Neurosci* 17: 489–517.

Semple DL, Anderson DJ (1992) Isolation of a stem cell for neurons and glia from the mammalian neural crest. *Cell* 71: 973–985.

Severson JA, Marcusson J, Winblad B, Finch CE (1982) Age-correlated loss of dopaminergic binding sites in human basal ganglia. *J Neurochem* 39: 1623–1631.

Shackelford TK, Larsen RG (1997) Facial asymmetry as an indicator of psychological, emotional, and physiological distress. *J Pers Soc Psychol* 72: 456–466.

Shadlen MF, Larson EB (1999) What's new in Alzheimer's disease treatment? Reasons for optimism about future pharmacologic options. *Postgrad Med* 105: 109–118.

Shanley DP, Kirkwood TBL (1998) Is the extended lifespan of calorie restricted animals an evolutionary adaptation? *Lifespan* 7: 3–5.

Sharp DH, Reinitz J (1998) Prediction of mutant expression patterns using gene circuits. *Biosystems* 47: 79–90.

Sherrington R, Froelich S, Sorbi S, Campion D, Chi H, Rogaeva EA, Levesque G, Rogaev EL, Lin C, Liang Y, Ikeda M, Mar L, Brice A, Agid Y, Percy ME, Clerget-Darpoux F, Piacentini S, Marcon G, Nacmias B, Amaducci L, Frebourg T, Lannfelt L, Rommens JM, St George-Hyslop PH (1996) Alzheimer's disease associated with mutations in presenilin 2 is rare and variably penetrant. *Hum Mol Genet* 5: 985–988.

Sherry DF, Galef BG Jr, Clark MM (1996) Sex and intrauterine position influence the size of the gerbil hippocampus. *Physiol Behav* 60: 1491–1494.

Sherwin BB (1998) Estrogen and cognitive functioning in women. *Proc Soc Exp Biol Med* 217: 17–22.

Sherwin BB, Tulandi T (1996) "Add-back" estrogen reverses cognitive deficits induced by a gonadotropin-releasing hormone agonist in women with leiomyomata uteri. *J Clin Endocrinol Metab* 81: 2545–2549.

Short RV (1977) The discovery of the ovaries. In: *The Ovary*, 2nd ed. (ed. Zuckerman S, Weir BJ), pp 1–39, Academic Press, New York.

Siegel MI, Doyle WJ (1975) Stress and fluctuating limb asymmetry in various species of rodents. *Growth* 39: 363–369.

Siegel MI, Doyle WJ, Kelley C (1977) Heat stress, fluctuating asymmetry, and prenatal selection in the laboratory rat. *Am J Phys Anthropol* 46: 121–126.

Sieweke MH, Graf T (1998) A transcription factor party during cell differentiation. *Curr Opinion Genet Dev* 8: 545–551.

Šimić G, Kostović I, Winblad B, Bogdanović N (1997) Volume and number of neurons of the human hippocampal formation in normal aging and Alzheimer's disease. *J Comp Neurol* 379: 482–494.

Simpson P, Woehl R, Usui K (1999) The development and evolution of bristle patterns in Diptera. *Development* 126: 1349–1364.

Sinclair D, Mills K, Guarente L (1998) Aging in *Saccharomyces cerevisiae*. *Annu Rev Microbiol* 52: 533–560.

Skeath JB, Doe CQ (1998) Sanpodo and Notch act in opposition to Numb to distinguish sibling neuron fates in the Drosophila CNS. *Development* 125: 1857–1865.

Slagboom PE, Vijg J (1989) Genetic instability and aging—theories, facts, and future perspectives. *Genome* 31: 373–385.

Slagboom PE, Droog S, Boomsma DI (1994) Genetic determination of telomere size in humans: a twin study of three age groups. *Am J Hum Genet* 55: 876–882.

Smith JR, Hayflick L (1974) Variation in the life-span of clones derived from human diploid cell strains. *J Cell Biol* 62: 48–53.

Smith JR, Pereira-Smith OM (1996) Replicative senescence: implications for in vivo aging and tumor suppression. *Science* 273: 63–67.

Smith JR, Whitney RG (1980) Intraclonal variation in proliferative potential of human diploid fibroblasts: stochastic mechanism for cellular aging. *Science* 207: 82–84.

Smith SC (1996) Pattern formation in the urodele mechanoreceptive lateral line: what features can be exploited for the study of development and evolution? *Int J Dev Biol* 40: 727–733.

Snieder H, MacGregor AJ, Spector TD (1998) Genes control the cessation of a woman's reproductive life: a twin study of hysterectomy and age at menopause. *J Clin Endocrinol Metab* 83: 1875–1880.

Sobel E, Louhija J, Sulkava R, Davanipour Z, Kontula K, Miettinen H, Tikkanen M, Kainulainen K, Tilvis R (1995) Lack of association of apolipoprotein E allele epsilon 4 with late-onset Alzheimer's disease among Finnish centenarians. *Neurology* 45: 903–907.

Sommer RJ, Sternberg PW (1995) Evolution of cell lineage and pattern formation in the vulval equivalence group of rhabditid nematodes. *Dev Biol* 167: 61–74.

Soong NW, Hinton DR, Cortopassi G, Arnheim N (1992) Mosaicism for a specific somatic mitochondrial DNA mutation in adult human brain. *Nat Genet* 2: 318–323.

Soulé M (1982) Allomeric variation. 1. The theory and some consequences. *Am Nat* 120: 751–764.

Soulé M, Cuzin-Roudy J (1982) Allomeric variation. 2. Developmental instability of extreme phenotypes. *Am Nat* 120: 765–786.

Spickett SG (1963) Genetic and developmental studies of a quantitative character. *Nature* 189: 870–873.

Spradling AC (1993) Developmental genetics of oogenesis. In: *The Development of Drosophila melanogaster* (ed. Bate M, Arias AM), pp 1–70, Cold Spring Harbor Laboratory Press, Cold Spring Harbor, NY.

Spreafico R, Frassoni C, Arcelli P, Selvaggio M, De Biasi S (1995) In situ labeling of apoptotic cell death in the cerebral cortex and thalamus of rats during development. *J Comp Neurol* 363: 281–295.

Stampfer MJ, Colditz GA, Willett WC, Manson JE, Rosner B, Speizer FE, Hennekens CH (1991) Postmenopausal estrogen therapy and cardiovascular disease. Ten-year follow-up from the nurses' health study. *N Engl J Med* 325: 756–762.

Stanner SA, Bulmer K, Andrés C, Lantseva OE, Borodina V, Poteen VV, Yudkin JS (1997) Does malnutrition in utero determine diabetes and coronary heart disease in adulthood? Results from the Leningrad siege study, a cross sectional study. *Br Med J* 315: 1342–1349.

Stary HC (1990) The sequence of cell and matrix changes in atherosclerotic lesions of coronary arteries in the first forty years of life. *Eur Heart J* 11 [Suppl E]: 3–19.

Stary HC (1994) Changes in components and structure of atherosclerotic lesions developing from childhood to middle age in coronary arteries. *Basic Res Cardiol* 89 [Suppl 1]: 17–32.

Stearns SC (1992) *The Evolution of Life Histories*, Oxford University Press, Oxford.

Steinmetz H (1996) Structure, functional and cerebral asymmetry: in vivo morphometry of the planum temporale. *Neurosci Biobehav Rev* 20: 587–591.

Steinmetz H, Herzog A, Huang Y, Häcklander T (1994) Discordant brain-surface anatomy in monozygotic twins. *N Eng J Med* 331: 952–953.

Steinmetz H, Herzog A, Schlaug G, Huang Y, Jänck L (1995) Brain asymmetry in monozygotic twins. *Cerebr Cortex* 5: 296–300.

Sternberg PW, Horvitz HR (1981) Gonadal cell lineages of the nematode *Panagrellus redivivus* and implications for evolution by the modification of cell lineage. *Dev Biol* 88: 147–266.

Stewart RJ, Preece RF, Sheppard HG (1975) Twelve generations of marginal protein deficiency. *Br J Nutr* 33: 233–253.

Stone LS (1933) The development of lateral-line sense organs in amphibians observed in living and vital-stained preparations. *J Comp Neurol* 57: 507–540.

Storrs EE, Williams RJ (1968) A study of monozygous armadillos in relation to mammalian inheritance. *Proc Natl Acad Sci USA* 60: 910–914.

Strom RC, Williams RH (1998) Cell production and cell death in the generation of variation in neuron number. *J Neurosci* 18: 9948–9953.

Suda J, Suda T, Ogawa M (1984) Analysis of differentiation of mouse hemopoietic stem cells in culture by sequential replating of paired progenitors. *Blood* 64: 393–399.

Supp DM, Witte DP, Potter SS, Brueckner M (1997) Mutation of an axonemal dynein affects left–right asymmetry in inversus viscerum mice. *Nature* 389: 963–936.

Swaab DF, Hofman MA (1990) An enlarged suprachiasmatic nucleus in homosexual men. *Brain Res* 537: 141–148.

Swaab DF, Gooren LJG, Hofman MA (1992) The human hypothalamus in relation to gender and sexual orientation. *Progr Brain Res* 93: 205–219.

Swaab DF, Zhou JN, Ehlhart T, Hofman MA (1994) Development of vasoactive intestinal polypeptide neurons in the human suprachiasmatic nucleus in relation to birth and sex. *Brain Res Dev Brain Res* 179: 249–259.

Swaddle JP, Cuthill IC (1995) Asymmetry and human facial attractiveness: symmetry may not always be beautiful. *Proc R Soc Lond [Biol]* 261: 111–116.

Swaddle JP, Witter MS (1994) Food, feathers, and fluctuating asymmetries. *Proc R Soc Lond [Biol]* 255: 147–152.

Swain A, Narvaez V, Burgoyne P, Camerino G, Lovell-Badge R (1998) Dax 1 antagonizes Sry action in mammalian sex determination. *Nature* 391: 761–767.

Swerdloff RS, Wang C, Hins M, Gorski RA (1992) Effect of androgens on the brain and other organs during development and aging. *Psychoneuroendocrinology* 17: 375–383.

Szilard L (1959) On the nature of the aging process. *Proc Natl Acad Sci USA* 45: 35–45.

Takeda T, Matsushita T, Kurozumi M, Takemura K, Higuchi K, Hosokawa M (1997) Pathobiology of the senescence-accelerated mouse (SAM). *Exp Gerontol* 32: 117–127.

Talman WT, Snyder D, Reis DJ (1980) Chronic lability of arterial pressure produced by destruction of A2 catecholamine neurons in rat brain stem. *Circ Res* 46: 842–853.

Tam PPL, Snow MHL (1981) Proliferation and migration of primordial germ cells during compensatory growth in mouse embryos. *J Embryol Exp Morphol* 64: 133–147.

Tang MX, Jacobs D, Stern Y, Marder K, Schofield P, Gurland B, Andrews H, Mayeux R (1996) Effect of oestrogen during menopause on risk and age at onset of Alzheimer's disease. *Lancet* 348: 429–432.

Tarani L, Lampariello S, Raguso G, Colloridi F, Pucarelli I, Pasquino AM, Bruni LA (1998) Pregnancy in patients with Turner's syndrome: six new cases and review of literature. *Gynecol Endocrinol* 12: 83–87.

Tease C, Fisher G (1986) Oocytes from young and old mice respond differently to colchicine. *Mutat Res* 173: 31–34.

Technau GM (1984) Fiber number in the mushroom bodies of adult *Drosophila melanogaster* depends on age, sex and experience. *J Neurogenet* 1: 113–126.

Teller JK, Russo C, DeBusk LM, Anelini G, Zaccheo D, Dagna-Bricarelli F, Scartezzini P, Bertolini S, Mann DM, Tabaton M, Gambetti P (1996) Presence of soluble amyloid beta-peptide precedes amyloid plaque formation in Down's syndrome. *Nature Med* 2: 93–95.

Tenniswood M (1986) Role of epithelial-stromal interactions in the control of gene expression in the prostate: an hypothesis. *Prostate* 9: 375–385.

Ter-Minassian M, Kowall NW, McKee AC (1992) Beta amyloid protein immunoreactive senile plaques in infantile Down's syndrome. *Soc Neurosci Abstr* 18: (#305.18), p 734.

Thomson JA, Itskovitz-Eldor J, Shapiro SS, Waknitz MA, Swiergiel JJ, Marshall VS, Jones JM (1998) Embryonic stem cell lines derived from human blastocysts. *Science* 282: 1145–1147.

Thornberry NA (1998) Caspases: key mediators of apoptosis. *Chem Biol* 5: R97–R103.

Thornhill R, Gangestad SW (1993) Human facial beauty: averageness, symmetry and parasite resistance. *Hum Nat* 4: 237–269.

Thornhill R, Gangestad SW, Comer R (1995) Human female orgasm and male fluctuating asymmetry. *Animal Behav* 50: 1601–1615.

Thummel CS, Burtis KC, Hogness DS (1990) Spatial and temporal patterns of E74 transcription during *Drosophila* development. *Cell* 61: 101–111.

Till JE, McCulloch EA (1961) A direct measurement of the radiation sensitivity of normal bone marrow cells. *Rad Res* 14: 213–222.

Till JE, McCulloch EA, Siminovich L (1964) A stochastic model of stem cell proliferation based on the growth of spleen colony forming cells. *Proc Natl Acad Sci USA* 51: 29–36.

Timms BG (1997) Anatomical perspective of prostate development. In: *Prostate: Basic and Clinical Aspects* (ed. Naz RK), pp 29–52, CRC Press, Boca Raton, FL.

Tissenbaum HA, Ruvkun G (1998) An insulin-like signaling pathway affects both longevity and reproduction in *Caenorhabditis elegans*. *Genetics* 148: 703–717.

Tonon L, Bergamaschi G, Dellavecchia C, Rosti V, Lucotti C, Malabarba L, Novella A, Vercesi E, Frassoni F, Cazzola M (1998) Unbalanced X-chromosome inactivation in haemopoietic cells from normal women. *Br J Haematol* 102: 996–1003.

Tononi G, Edelman GM (1998) Consciousness and complexity. *Science* 282: 1846–1851.

Torgerson DJ, Thomas RE, Reid DM (1997) Mothers and daughters menopausal ages: is there a link? *Eur J Obstet Gynecol Reprod Biol* 74: 63–66.

Torigoe C, Inman JK, Metzger H (1998) An unusual mechanism for ligand antagonism. *Science* 281: 568–572.

Trainor KJ, Wigmore DJ, Chrysostomou A, Dempsey JL, Seshadri R, Morley AA (1984) Mutation frequency of human lymphocytes increases with age. *Mech Ageing Dev* 27: 83–86.

Tramo MJ, Loftus WC, Thojas CE, Green RL, Mott LA, Gazzaniga MS (1995) Surface area of human cerebral cortex and its gross morphological subdivisions: in vitro measurements in monozygotic twins suggest differential hemispheric effects of genetic factors. *J Cogn Neurosci* 7: 292–301.

Tramontin AD, Smith GT, Breuner CW, Brenowitz EA (1998) Seasonal plasticity and sexual dismorphism in the avian song control system: stereological measurement of neuron density and number. *J Comp Neurol* 396: 186–192.

Treloar AE, Boynton RE, Behn BG, Brown BW (1967) Variation of the humam menstrual cycle through reproductive life. *Int J Fertil* 12: 77–126.

Treloar AE (1981) Menstrual cyclicity and the pre-menopause. *Maturitas* 3: 162–172.

Treloar SA, Do KA, Martin NG (1998a) Genetic influences on the age at menopause. *Lancet* 352: 1084–1085.

Treloar SA, Martin NG, Heath AC (1998b) Longitudinal genetic analysis of menstrual flow, pain, and limitation in a sample of Australian twins. *Behav Genet* 28: 107–116.

Trivers RL, Willard DE (1973) Natural selection of parental ability to vary the sex ratio of offspring. *Science* 179: 90–92.

Tsuji K, Nakahata T (1989) Stochastic model of multipotent hemopoietic progenitor differentiation. *J Cell Physiol* 139: 647–653.

Turing AM (1950) The chemical basis of morphogenesis. *Philos Tran R Soc [Biol]* 237: 37–72.

Tyrrell J, Cosgrave M, Hawi Z, McPherson J, O'Brien C, McCalvert J, McLaughlin M, Lawlor B, Gill M (1998) A protective effect of apolipoprotein E e2 allele on dementia in Down's syndrome. *Biol Psychiatry* 43: 397–400.

Uemura K, Pisa Z (1985) Recent trends in cardiovascular mortality in industrialised countries. *World Health Stat Q* 38: 142–162.

Uitdehaag BM, Polman CH, Valk J, Koetsier JC, Lucas CJ (1989) Magnetic resonance imaging in multiple sclerosis twins. *J Neurol Neurosurg Psychiatry* 52: 1417–1419.

Upadhyay S, Zamboni L (1982) Ectopic germ cells: natural model for the study of germ cell sexual differentiation. *Proc Natl Acad Sci USA* 79: 6584–6588.

Usui K, Kimura K-I (1993) Sequential emergence of the evenly spaced microchaetes on the notum of *Drosophila*. *Roux's Arch Devel Biol* 203: 151–158.

van Assche FA, Aerts L (1985) Long-term effect of diabetes and pregnancy in the rat. *Diabetes* 34 [Suppl 2]: 116–118.

van Beijsterveldt CEM, Molenaar PCM, de Geus EJC, Boomsma DI (1996) Heritability of human brain functioning as assessed by electroencephalography. *Am J Hum Genet* 58: 562–573.

Vandenbergh JG, Huggett CL (1994) Mother's prior intrauterine position affects the sex ratio of her offspring in house mice. *Proc Natl Acad Sci USA* 91: 11055–11059.

Van der Put NM, Vanden-Heuvel LP, Steegers-Theunissen RPM, Trijbels F-J, Eskes TK, Mariman EC, den Heyer M, Blom HJ (1995) Decreased methylene tetrahydrofolate reductase activity due to the 677C T mutation in families with spina bifida offspring. *J Mol Med* 74: 691–694.

van Leeuwen FW, Burbach JP, Hol EM (1998) Mutations in RNA: a first example of molecular misreading in Alzheimer's disease. *Trends Neurosci* 21: 331–335.

Van Valen L (1962) A study of fluctuating asymmetry. *Evolution* 16: 125–142.

Vaupel JW, Carey JR (1993) Compositional interpretations of medfly mortality. *Science* 260: 1666–1667.

Vaupel JW, Carey JR, Christensen K, Johnson TE, Yashin AI, Holm NV, Iachine IA, Kannisto V, Khazaeli AA, Liedo P, Longo VD, Zeng Y, Manton KG, Curtsinger JW (1998) Biodemographic trajectories of longevity. *Science* 280: 855–860.

Veneman TF, Beverstock GC, Exalto N, Mollevanger P (1991) Premature menopause because of an inherited deletion in the long arm of the X-chromosome. *Fertil Steril* 55: 631–633.

Verhoeven G, Hoeben E (1994) Pathogenesis of benign prostatic hyperplasia: potential role of mesenchymal-epithelial interactions. In: *Benign Prostatic Hyperplasia. Innovations in Management*, pp 17–37, Springer-Verlag, Berlin.

Vogel F, Motulsky AG (1997) *Human Genetics: Problems and Approaches*, 3rd ed., Springer-Verlag, Berlin.

vom Saal FS (1981) Variation in phenotype due to intrauterine positioning of male and female fetuses in rodents. *J Repro Fertil* 62: 633–650.

vom Saal FS (1983) The interactions of circulating oestrogens and androgens in regulating mammalian sexual orientation. In: *Hormones and Behavior in Higher Vertebrates*, J. Bathalzart, E, Pröve, R Gilles (eds), pp 159–177, Springer-Verlag, Berlin.

vom Saal FS (1984) The intrauterine position phenomenon: effects on physiology, aggressive behavior, and population dynamics in house mice. In: *Biological Perspectives on Aggression* (ed. Flannelly K, Blanchard R, Blanchard D), pp 135–179, Liss, New York.

vom Saal FS (1989) Sexual differentiation in litter bearing mammals: influence of sex of adjacent fetuses in utero. *J Anim Sci* 67: 1824–1840.

vom Saal FS, Bronson FH (1978) In utero proximity of female mouse fetuses to males: effect on reproductive performance during later life. *Biol Reprod* 19: 842–853.

vom Saal FS, Bronson FH (1980) Variation in length of the estrous cycle in mice due to former intrauterine proximity to male fetuses. *Biol Reprod* 22: 777–780.

vom Saal FS, Dhar MD (1992) Blood flow in the uterine loop artery and loop vein is bidirectional in the mouse: implications for intrauterine transport of steroids. *Physiol Behav* 52: 163–171.

vom Saal FS, Moyer CL (1985) Prenatal effects on reproductive capacity during aging in female mice. *Biol Reprod* 32: 1116–1126.

vom Saal FS, Quadagno DM, Even MD, Keisler LW, Keisler DH, Khan S (1990) Paradoxical effects of maternal stress on fetal steroids and postnatal reproductive traits in female mice from different intrauterine positions. *Biol Reprod* 43: 751–761.

vom Saal FS, Nelson JF, Finch CE (1994) The natural history of reproductive humans, laboratory rodents and selected other vertebrates. In: *Physiology of Reproduction* (ed. Knobil E), pp 1213–1314, Raven Press, New York.

vom Saal FS, Timms BG, Montano MM, Palanza P, Thayer KA, Nagel SC, Dhar MD, Ganjam VK, Parmigiani S, Welshons WV (1997) Prostate enlargement in mice due to fetal exposure to low dosages of estradiol or diethylstilbestrol and opposite effects at high dosages. *Proc Natl Acad Sci USA* 94: 2056–2061.

vom Saal FS, Clark MM, Galef BG Jr, Drickamer LC, VandenBergh JG (1998a) Intrauterine position phenomenon. In: *Encyclopedia of Reproduction*, art 231, vol 2, pp113–120, Academic Press, San Diego.

vom Saal FS, Welshons WV, Hansen LG (1998b) Organochlorine residues and breast cancer. *N Engl J Med* 338: 988 [discussion 991].

von Boehmer H (1997) T-cell development: is Notch a key player in lineage decisions? *Curr Biol* 7: R308–R310.

Vuillaume I, Vermersch P, Destée A, Petit H, Sablonnière B (1998) Genetic polymorphisms adjacent to the CAG repeat influence clinical features at onset in Huntington's disease. *J Neurol Neurosurg Psychiatry* 64: 758–762.

Wachter K, Finch CE (eds.) (1997) *Biodemography of Aging*, National Academy of Sciences, Washington, D.C.

Waddington CH (1960) Experiments on canalizing selection. *Genet Res Cambridge* 1: 140–150.

Wade MG, Desaulniers D, Leingartner K, Foster WG (1997) Interactions between endosulfan and dieldrin on estrogen-mediated processes in vitro and in vivo. *Reprod Toxicol* 11: 791–798.

Wagner CK, Nakayama AY, De Vries GJ (1998) Potential role of maternal progesterone in the sexual differentiation of the brain. *Endocrinology* 139: 3658–3661.

Wakamatsu Y, Maynard TM, Jones SU, Weston JA (1999) NUMB localizes in the basal cortex of mitotic avian neuroepithelial cells and modulates neuronal differentiation by binding to NOTCH-1. *Neuron* 23: 71–81.

Wakayama T, Perry ACF, Zucoti M, Johnson KR, Yanagimachi R (1998) Full-term development of mice from enucleated oocytes injected with cumulus cell nuclei. *Nature* 394: 369–374.

Walsh C, Cepko CL (1992) Widespread dispersion of neuronal clones across functional regions of the cerebral cortex. *Science* 255: 434–440.

Walsh C, Cepko CL (1993) Clonal dispersion in proliferative layers of developing cerebral cortex. *Nature* 362: 632–635.

Walsh C, Cepko CL, Ryder EF, Church GM, Tabin C (1992) The dispersion of neuronal clones across the cerebral cortex. *Science* 258: 317–320.

Wang X, DeKosky ST, Wisniewski S, Aston CE, Kamboh MI (1998) Genetic association of two chromosome 14 genes (presenilin 1 and alpha 1-antichymotrypsin) with Alzheimer's disease. *Ann Neurol* 44: 387–390.

Ward BC, Noordeen EJ, Nordeen KW (1998) Individual variation in neuron number predicts differences in the propensity for avian vocal imitation. *Proc Natl Acad Sci USA* 95: 1277–1282.

Ward S, Carrel JS (1979) Fertilization and sperm competition in the nematode *Caenorhabditis elegans*. *Dev Biol* 73: 304–321.

Ward S, Thomson N, White JG, Brenner S (1975) Electron microscopical reconstruction of the anterior sensory anatomy of the nematode *Caenorhabditis elegans*. *J Comp Neurol* 160: 313–337.

Waterland RA, Garza C (1999) Potential mechanisms of metabolic imprinting that lead to chronic disease. *Am J Clin Nutr* 69: 179–197.

Weindruch RH, Walford R (1998) *The Retardation of Aging and Disease by Dietary Restriction*, Thomas, Springfield, Ill.

Weinmaster G (1997) The ins and outs of Notch signaling. *Mol Cell Neurosci* 9: 91–102.

Weisgraber KH (1994) Apolipoprotein E: structure-function relationships. *Adv Protein Chem* 45: 249–302.

Weiss S, Uematsu Y, D'Hoostelaere L, Alanen A (1989) Gross genetic differences among substrains of NZB mice. *J Immunogenet* 16: 217–221.

West MJ (1999) Stereological methods for estimating the total number of neurons and synapses: issues of precision and bias. *Trends Neurosci* 22: 51–61.

Weisz J, Ward IL (1980) Plasma testosterone and progesterone titers of pregnant rats, their male and female fetuses, and neonatal offspring. *Endocrinology* 106: 306–316.

Welker C (1976) Receptive fields of barrels in the somatosensory neocortex of the rat. *J Comp Neurol* 166: 173–190.

Weng NP, Granger L, Hodes RJ (1997) Telomere lengthening and telomerase activation during human B cell differentiation. *Proc Natl Acad Sci USA* 94: 10827–10832.

Weng NP, Hathcock KS, Hodes RJ (1998) Regulation of telomere length and telomerase in T and B cells: a mechanism for maintaining replicative potential. *Immunity* 9: 151–157.

Werle E, Fiehn W, Hasslacher C (1998) Apolipoprotein E polymorphism and renal function in German type 1 and type 2 diabetic patients. *Diabetes Care* 21: 994–998.

West MJ (1993) Regionally specific loss of neurons in the aging human hippocampus. *Neurobiol Aging* 14: 287–293.

West MJ, Slimianka L, Gundersen HJG (1991) Unbiased stereological estimation of the total number of neurons in the subdivisions of the rat hippocampus using the optical fractionator. *Anat Rec* 231: 482–497.

Weverling-Rijnsburger AWE, Blaum GJ, Lagaay AM, Knook DL, Meinders AE, Westendorp RGJ (1997) Total cholesterol and risk of mortality in the oldest old. *Lancet* 350: 1119–1123.

Whitaker HA, Selnes OA (1975) Anatomic variations in the cortex. Individual differences and the problem of the localization of language functions. *Ann NY Acad Sci* 280: 844–834.

White JG, Southgate E, Thomson JN, Brenner S (1976) The structure of the ventral nerve cord of *Caenorhabditis elegans*. *Philos Trans R Soc Lond [Biol]* 275: 327–348.

Wickelgren I (1996) For the cortex, neuron loss may be less than thought. *Science* 273: 48–50.

Wierda M, Goudsmit E, Van der Woude, Purba JS, Hofman MA, Bogte H, Swaab DF (1991) Oxytocin cell number in the human paraventricular nucleus remains constant with aging and in Alzheimer's disease. *Neurobiol Aging* 12: 511–516.

Williams GC (1957) Pleiotropy, natural selection and the evolution of senescence. *Evolution* 11: 398–411.

Williams RW, Herrup K (1988) The control of neuron number. *Annu Rev Neurosci* 11: 423–453.

Williams RW, Rakic P (1988) Three-dimensional counting: an accurate and direct method to estimate numbers of cells in sectioned material. *J Comp Neurol* 278: 344–352.

Williams RW, Strom RC, Rice DS, Goldowitz D (1996) Genetic and environmental control of variation in retinal ganglion cell number in mice. *J Neurosci* 16: 7193–7205.

Williams RW, Strom RC, Goldowitz D (1998) Natural variation in neuron number in mice is linked to a major quantitative trait locus on Chromosome 11. *J Neurosci* 18: 138–146.

Wilmut I, Schnieke AE, McWhir J, Kind AJ, Campbell KH (1997) Viable offspring derived from fetal and adult mammalian cells. *Nature* 385: 810–813.

Wingfield JC, Hegner RE, Dufty JR, Ball GF (1990) The "challenge hypothesis": theoretical implications for patterns of testosterone secretion, mating systems, and breeding strategies. *Am Nat* 136: 829–846.

Winklbauer R (1988) Growth control and pattern regulation in the lateral line system of *Xenopus*. *Z Naturforsch* 43c: 294–300.

Winklbauer R (1989) Development of the lateral line system in *Xenopus*. *Progr Neurobiol* 32: 181–206.

Winklbauer R, Hausen P (1983a) Development of the lateral line system in *Xenopus laevis*. I. Normal development and cell movement in the supraorbital system. *J Embryol Exp Morphol* 76: 265–281.

Winklbauer R, Hausen P (1983b) Development of the lateral line system in *Xenopus laevis*. II. Cell multiplication and organ formation in the supraorbital system. *J Embryol Exp Morphol* 76: 283–296.

Winklbauer R, Hausen P (1985a) Development of the lateral line system in *Xenopus laevis*. III. Development of the supraorbital system in triploid embryos and larvae. *J Embryol Exp Morphol* 88: 183–192.

Winklbauer R, Hausen P (1985b) Development of the lateral line system in *Xenopus laevis*. IV. Pattern formation in the supraorbital system. *J Embryol Exp Morphol* 88: 193–207.

Winton DJ, Ponder BAJ (1990) Stem-cell organization in mouse small intestine. *Proc R Soc Lond [Biol]* 241: 13–18.

Winton DJ, Blount MA, Ponder BAJ (1988) A clonal marker induced by mutation in mouse intestinal epithelium. *Nature* 333: 463–466.

Wise PM, Krajnak KM, Kashon ML (1996) Menopause: the aging of multiple pacemakers. *Science* 273: 67–70.

Withers HR, Elkind MM (1970) Microcolony survival assay for cells of mouse intestinal mucosa exposed to radiation. *Int J Radiat Biol* 17: 261–268.

Wolff JM, Boeckmann W, Mattalaer P, Handt S, Adam G, Jakse G (1995) Determination of prostate gland volume by transrectal ultrasound: correlation with radical prostatectomy specimens. *Eur Urol* 28: 10–12.

Wolpert L, Beddington R, Brockes J, Jessel T, Lawrence P, Meyerowitz E (1998) *Principles of Development*, Oxford University Press, Oxford.

Wong-Riley MTT, Welt C (1980) Histochemical changes in cytochrome oxidase of cortical barrels after vibrissal removal in neonatal and adult mice. *Proc Natl Acad Sci USA* 77: 2333–2337.

Wood WB (1991) Evidence from reversal of handedness in *C. elegans* embryos for early cell interactions determining cell fates. *Nature* 349: 536–538.

Wright S (1920) The relative importance of heredity and environment in determining the piebald pattern of guinea pigs. *Proc Natl Acad Sci USA* 6: 321–332.

Wyllie AH, Kerr JFR, Currie AR (1980) Cell death: the significance of apoptosis. *Int Rev Cytol* 68: 251–306.

Yaffe K, Sawaya G, Lieberburg I, Grady D (1998) Estrogen therapy in postmenopausal women: effects on cognitive function and dementia. *JAMA* 279: 688–695.

Yashin AI, Manton KG, Woodbury MA, Stallard E (1995) The effects of health histories on stochastic process models of aging and mortality. *J Math Biol* 34: 1–16.

Ye L, Miki T, Nakura J, Oshima J, Kamino K, Rakugi H, Ikegami H, Higaki J, Edland SD, Martin GM, Ogihara T (1997) Association of a polymorphic variant of the Werner helicase gene with myocardial infarction in a Japanese population. *Am J Med Genet* 68: 494–498.

Ye Y, Lukinova N, Fortini ME (1999) Neurogenic phenotypes and altered Notch processing in *Drosophila* Presenilin mutants. *Nature* 398: 525–529.

Yokoyama T, Copeland NG, Jenkins NA, Montgomery CA, Elder FF, Overbeek PA (1993) Reversal of left–right symmetry: a situs invertus mutation. *Science* 260: 679–682.

Yost HJ (1992) Regulation of vertebrate left–right asymmetries by extracellular matrix. *Nature* 357: 158–161.

Young BK, Suidan J, Antoine C, Silverman F, Lustig I, Wasserman J (1985) Differences in twins: the importance of birth order. *Am J Obstet Gynecol* 151: 915–921.

Yu CE, Oshima J, Fu YH, Wijsman EM, Hisama F, Alisch R, Matthews S, Nakura J, Miki T, Ouais S, Martin GM, Mulligan J, Schellenberg GD (1996) Positional cloning of the Werner's syndrome gene. *Science* 272: 258–262.

Yuh CH, Bolouri H, Davidson EH (1998) Genomic cis-regulatory logic: experimental and computational analysis of a sea urchin gene. *Science* 279: 1896–1902.

Zamenhoff S, van Marthens E, Grauel L (1971) DNA (cell number) in neonatal brain: second generation (F_2) alteration by maternal (F_0) dietary protein restriction. *Science* 172: 850–851.

Zavala C, Herner G, Fialkow PF (1977) Evidence for selection in cultured diploid fibroblast strains. *Exp Cell Res* 117: 137–144.

Zhang C, Lui VW, Addessi CL, Sheffield DA, Linnane AW, Nagley P (1998) Differential occurrence of mutations in mitochondrial DNA of human skeletal muscle during aging. *Hum Mutat* 11: 360–371.

Zhong W, Feder JN, Jiang M-M, Jan LY, Jan YN (1996) Asymmetric localization of a mammalian *Numb* homolog during mouse cortical neurogenesis. *Neuron* 17: 43–63.

Zhou JN, Hofman MA, Gooren LJ, Swaab DF (1995) A sex difference in the human brain and its relation to transsexuality. *Nature* 378: 68–70.

Zondek T, Zondek LH (1975) The fetal and neonatal prostate. In: *Normal and Abnormal Growth of the Prostate*, pp 5–28, Thomas, Springfield, Ill.

Subject Index

259

Author Index